Under Ancient Skies

Under Ancient Skies

Ancient Astronomy and Terrestrial Catastrophism

Paul Dunbavin

Third Millennium Publishing

First published in 2005 by Third Millennium Publishing
PO Box 20, Long Eaton, Nottingham NG10 5BN
www.third-millennium.co.uk

British Library Cataloguing in Publication Data
A catalogue record for this book is available from the British Library

ISBN 0-9525029-2-5

Prepared, printed and distributed by:
York Publishing Services Ltd
64 Hallfield Road
Layerthorpe
York YO31 7ZQ
Tel: 01904 431213; Website: www.yps-publishing.co.uk

Contents

Introduction

The true astronomer must be the wisest of men, understanding
by the term…the man who investigates the seven revolutions
included in the eight revolutions [i.e. the revolutions of the
sun, the moon, and the five planets, apart from the eighth
motion, that of the daily rotation] each of the seven describing
its own circle in a manner such as would never be easily
comprehended by any one unless he possessed extraordinary
powers.
Plato, Epinomis, 99 A, B. (translation by Sir Thomas Heath)

As an amateur astronomer, the possibility that the Earth has experienced
a geological cataclysm during the period of human prehistory is a question
that has intrigued me from an early age. Astronomers appreciate that the
universe is a violent place and that, from time to time, forces beyond our
earthly comprehension may visit us from the depths of space. Such an
event was surely Noah's Flood. From the viewpoint of an astronomer, it
seems to me that the Great Flood could only have had an astronomical
cause; and that if we search in the right places there may be astronomical
proof of its reality.

In 1995, I published *The Atlantis Researches – the Earth's Rotation in
Mythology and Prehistory*, which represented the results of my research up
to 1992. Extrapolating from a pattern of evidence, the theory proposed
a comet impact during Neolithic times, towards the close of the fourth
millennium BC. This event was remembered world-wide in myths and
legends and could also be detected in a pattern of physical evidence. It
triggered a change in the characteristics of the daily rotation, permanently
altering world climate and sea levels – yet it was just the most recent of
many such impacts that the Earth has endured.

The majority of comment at the time concerned the search for Lost
Atlantis rather than the wider consequences of such a cataclysm; for it is

fair to say that *The Atlantis Researches* has rendered obsolete much of the earlier discussion on that subject. Many readers skipped over those parts of the hypothesis that fell outside their own knowledge or interest.

For the study of catastrophism in human prehistory is not a subject for the academic specialist. It remains a multi-disciplinary problem. There is just not enough evidence within the bounds of any single discipline to make a case one way or the other. There can be no 'experts' on catastrophism, nor any professors of multidisciplinary studies. To make a case for the Flood as a real prehistoric catastrophe, one must be prepared to review evidence from disciplines as varied as: mythology, folklore, archaeology, Egyptology, geology, geophysics, astronomy, climatology, sea-level research, ancient history, etymology, calendars, horology, meteorology, oceanography… and probably many other disciplines and sub-disciplines. Absurd then, to think that a research paper could be refereed in the academic way by say, an archaeologist or a geologist – yet this seems to be what is expected! In any case, which is the proper journal to publish in? And on which library shelf should a book on the subject belong?

Even as recently as the early 1990's catastrophism remained a subject for the fringe. Eminent academics, if they favoured catastrophism, would approach the subject in only the most guarded terms. However, in that decade a whole series of important developments took place. These have rendered the study of catastrophist geology, and particularly the likelihood of a comet impact, a far more respectable pursuit.

For most of twentieth century, the extinction of the dinosaurs sixty-five million years ago was a hotly-debated subject. The theory that their extinction followed a giant asteroid impact was but one among many ideas proposed. The discovery of the Chicxulub impact crater in Yucatan has at last provided the 'hard evidence' that the catastrophist geologists needed. It has become respectable to discuss the theory that much of the geological time frame has been punctuated by impact events, with research papers now appearing in even the most prestigious journals.

However, the possibility of an asteroid or comet impact within more recent *human* prehistory remained too close to home for many people. The fall of comet Shoemaker-Levy-9 on Jupiter put a stop to all that and has probably saved twenty years of debate towards making the subject respectable. One simply has to say: 'Yes, a comet could hit the Earth today – just look what happened on Jupiter!'

This study will therefore probe the wider consequences of astronomical catastrophes in recent millennia; their effect on the Earth; and their importance in human prehistory. I have assumed throughout a threshold knowledge and interest in astronomy – only because the book has to be classified *somewhere*. Using the evidence of calendars and eclipses as well as other physical evidence, I set out to probe ancient history and mythology, hoping to discover *how* and *when* the Earth's rotation had been altered during the past ten thousand years. The workings of the real world are not always simple and cannot be structured like a work of fiction. However, I shall attempt to explain all the terms for the benefit of the cross-disciplinary reader. I hope you will find this to be a *real* book, with a new idea or suggestion in every chapter.

Many people, I know, will say that the daily rotation of the Earth and the appearances of comets and planets are a dull subject, of little interest to the average person. That view would change very rapidly if a comet were to strike the Earth today. Imagine yourself on the seashore on a bright sunny day looking out on a beautiful calm sea. Unknown to you a small energetic comet has just struck somewhere on the far side of the world and the Earth has begun to wobble on its axis. The natural level of the sea has just changed and now it wants to be somewhere above your head! At first, you may think the tide is coming in, but no, it's not the tide, nor even a tsunami. Look! The entire ocean is beginning to flow downhill towards you, as if it had been tipped out of some gigantic bowl.

It's already far too late to think of running – you may as well just watch!

1

Myths of a Better Time

From all over the world, we may find ancient traditions that the world was once destroyed in a catastrophic flood. One hesitates to call these 'myths', for one person's myth is another's religion and what for the former is an intriguing tale, is for the latter a matter of faith. Until the scientific revolution of recent centuries, few would have dared to question the reality of the Flood of Noah, but today we can take a more objective view and pose the question scientifically. Could the Flood story found in so many of our myths and religions, be a distant memory of a real ancient cataclysm?

Although the Judeo-Christian tradition would view the Flood as a punishment for the wickedness of mankind, there are other traditions too, that recall a better time in the distant past, a golden age before the flood, when life was easier. Could this too be a real memory of the ancient past?

The story of Noah's Flood is related in the Old Testament Book of Genesis and also in the Muslim Koran. We may presume that it was among the beliefs that Abraham and his followers took with them when they left Ur of the Chaldees (around 2300-2000 BC). The god of the Hebrews resolves to cleanse the corrupt world, but promises to spare Noah and his family, because he alone is a good man.

Noah is instructed to build an ark and take aboard a male and female of every creature. The ark is a true ship of gopher wood, securely sealed with pitch, and the Biblical account is precise in its dimensions: three hundred cubits, by fifty cubits and thirty cubits high.[1] There were three levels and many rooms; and a door through which the animals were led. This is typical of all the Mesopotamian versions, which offer more information about the ark than about the Flood itself.

1

We are told: "all the fountains of the great deep [were] broken up, and the windows of heaven were opened".[2] This is as close as we get to a cause or mechanism. According to Josephus, Adam, the first man, had predicted that the world would one day be destroyed by fire and water.[3] We may see that the floodwaters came both from below (the Flood) and from above (the Deluge). The rain falls continuously for forty days and nights and the entire world is submerged. The Flood is truly, universal. Noah sends out various birds in search of land but they circle and return; and a whole year of Noah's life passes before the waters abate enough for them to leave the ark.

Western science has neglected to analyse the story of the Flood with the rigour that it has devoted to other matters. There are historical reasons for this. Modern scientific thinking has emerged from a mainly European Christian background; and for centuries was forced to do battle with religious orthodoxy. The triumph of science in the nineteenth century was very much a defeat for the Biblical doctrine. The Deluge became a subject of interest only to ecclesiastical students and mythographers – neither of whom wield much influence with the scientific community.

The Flood ceased to be regarded as an ancient reality and as such, it no longer demanded any explanation by science. As a product of the human imagination, the similarities between the various flood myths from around the world could be dismissed as a kind of primitive fiction. To quote Edmund Sollberger, it could be explained as, "the product of the human mind's similar reaction to similar circumstances".[4]

Flood Myths from Ancient Mesopotamia

The Mesopotamian version of the Flood is the most likely inspiration for the Biblical Flood. It is found in the Epic of Gilgamesh, discovered among the ruins of ancient Nineveh, on cuneiform tablets dating from the second millennium BC. Its hero Gilgamesh was an early king of the Sumerian city of Uruk who, according to the king lists, reigned at about 2700 BC.[5] This assures us that the underlying reality of the Flood must date to centuries earlier than this.

Several generations after the Flood, Gilgamesh sets out on a quest for the secret of eternal life. He seeks Utnapishtim, a survivor of the Flood who dwells in a distant land "beyond the waters of death" – evidently the Sumerian paradise of the dead. He is transported across the ocean by the ferryman Urshanabi who leads him to the land of Dilmun, where he meets with Utnapishtim.

Utnapishtim relates how he survived the Great Flood. Formerly he had lived in the Sumerian city of Shurrupak. One day, the god Ea warned that En-lil was irritated by the incessant clamour of humankind and intended to destroy them all in a great flood. Ea gave him instructions to build a ship, which we may equate with the Biblical ark. The 'ark' of Utnapishtim however, was a strange craft indeed: a perfect cube of a hundred and twenty Babylonian cubits each side; seven stories high and seven compartments each side. When this strange ark floated, a third lay above the water and two thirds below. Utnapishtim loaded the vessel with animals, both wild and domesticated, and waited for the deluge to begin.

As the sky darkened, the gods themselves hid in terror and the torrential rain poured for seven days. The vessel floated away and came to rest on the slope of mount Nisir, east of the upper Tigris. After seven more days had elapsed Utnapishtim let out birds, a dove, a swallow and finally a raven, which found dry land and did not return; and so he felt safe to leave the ship.

When the gods saw that a few people had survived, they squabbled among themselves and Ishtar (Venus) begged Enlil to spare them. Enlil therefore relented and granted eternal life to Utnapishtim and his wife, on condition that they dwell far away "at the mouth of the rivers". Utnapishtim does reveal the secret of eternal life – a certain plant – but on the way home, a serpent rises from the sea and snatches it from Gilgamesh. Thus, the moral of the tale is that only the gods may live forever.

Berossus

Another Mesopotamian version of the Flood is given in Greek translation by Berossus in his Babyloniaca. He maintained that civilisation was the gift of the gods to mankind, and that the god Oannes had first transmitted this wisdom to humanity in Babylon, at a remote era some 432,000 years before the Flood. The reigns of ten divine kings spanned this incredibly long period, the last of whom was Xisouthros, in whose reign the Flood occurred.[6]

In the version of Berossus, the god Ea, or Enki, warns Xisouthros of the coming Flood in a dream. He translates Enki as the Greek god *Cronus*, the Roman *Saturn*. The justification for this seems to be that Berossus translated the Babylonian god Marduk as Zeus; and Enki was the father of Marduk, just as, in Greek mythology, Cronus was the father of Zeus. Marduk was also called En-lil, or Bel, the Baal of the Bible.

Ea tells Xisouthros to bury "the beginnings and the middles and the ends of all writings" in the city of Sippar – a common formula meaning simply "all".[7] Thus was knowledge of the world before the Flood preserved. Otherwise, the story is similar to the other accounts, even down to the details of the birds that are released. Berossus tells us that some ten generations after the Flood a great sage arose who recovered the ancient knowledge of astronomy; one presumes, from the buried records.

In 1914 an even older version of the Flood myth was discovered, on cuneiform tablets, at the site of the Sumerian city of Nippur.[8] It tells of a pious king named Ziusudra who, like Noah and Utnapishtim, was forewarned of the Flood. The Xisouthros of Berossus is seen to be a Hellenised version of the name Ziusudra of the Sumerian Flood. Moreover, his name occurs in the Babylonian king lists, where he is listed among the long-lived kings *before* the Flood. Since Ziusudra follows king Ubar-Tutu in the king lists, we may assume that he is one and the same as Utnapishtim, who is called 'son of Ubar-Tutu' in the Gilgamesh Epic.

We gain no further insight from any Babylonian source, as to the ultimate causes of the Flood; it is simply a punishment for the wickedness of mankind. Other mythological sources are more informative on this point and it may be that other Mesopotamian stories can supply this missing detail.

Deucalion and Pyrrha

In the equivalent Greek myths, the Flood was associated with Deucalion, son of the Titan Prometheus and his wife Pyrrha; and like the Babylonian version, the Flood is brought on by a deluge of rain.[9] Prometheus warns Deucalion to build a "chest" and provision it. Apollodorus describes the Flood thus:

> Zeus, by pouring heavy rain from heaven flooded the greater part of Greece, except a few who fled to the high mountains in the neighbourhood. It was then that the mountains in Thessaly parted, and that all the world outside the isthmus and Peloponnese was overwhelmed.[10]

The chest holding Deucalion (and presumably Pyrrha also) floats for nine days until the rain ceases and they land at Parnassus. The survival of the human race is explained by the little tale that Deucalion threw stones over his head, which became men, while those thrown by Pyrrha became women. Although we are told that a few other people have survived in the hills, there is no 'ark' to explain the survival of the animals.

Again, the Flood is a *deluge*. The rain comes not from clouds, but is simply poured from the sky, and yet there is also the sense of a great wave of water arriving from elsewhere to shatter the mountains and flood the low-lying regions.

In Greek mythology, the Titans were the divine dynasty, the Children of Uranus (sky) and Ge (earth). Scholars generally conclude that they symbolise the elementary forces of nature, but their true nature need not concern us here, beyond that required for further analysis of the flood story. The youngest son of Uranus and Ge was Cronus, who usurped his father's throne and ruled as a tyrant. He in turn was overthrown by his own son Zeus, who became the mightiest of the Greek gods.

The planets were named after Titans: Saturn was the star of Cronus and Jupiter was Zeus, Venus for Aphrodite, Mars for Aries and Mercury is Hermes. Today we use the equivalent Roman gods as names for the planets, but the association is truly ancient. Other Titans personified earthly phenomena, such as Poseidon the sea god and Hades god of the underworld. Thus we may see that Earth and Sky created all the planets and all the geography of the world; and could destroy it again if they so wished.

A parallel tradition recalled the reign of Cronus as an age of happiness for mankind and this is told in the Hesiod's myth of the ages.[11] In the Golden Age, men lived long happy lives free from toil and never aged, but eventually died in their sleep. The golden race was hidden in the ground where they persist as the spirits of the earth (the daemons). The Golden Age degenerated into the Silver Age of Zeus in which men's lives were brief. The men of the Silver Age were sinful and dishonoured the gods; and so Zeus consigned them to the underworld. They too became honoured spirits, to be replaced by the race of bronze, which went the same way. The honoured heroes of the Iliad followed them in their turn. Of Cronus, we are told that he became king of the Islands of the Blessed where the heroes of Troy were consigned after death. It is evidently another view of the land of the dead. May we see in this the gradual decay of a historical tradition into a timeless poetic myth?

According to the Roman poet Ovid, the Golden Age was a world of eternal springtime. The four seasons, as we know them, began only when Saturn (Cronus) fell to the dark underworld and the age of Zeus opened.[12] The transition was a time of great extremes in the weather and people took shelter in caves. Could this too, be a real memory of a time of climate change? If it were a reality then we would expect it to have a counterpart in other mythologies.

That the Flood *follows* the Golden Age of the Titans is evident from the genealogy of Deucalion. He was the son of Prometheus, and grandson of Iapetus, a brother of Cronus. Iapetus was among the Titans confined in the underworld after Zeus overthrew Cronus. A simplistic analysis therefore makes the Flood a generation later than the Golden Age.

Angered by this treatment of her children, Ge gave birth to the giants (gigantes). According to Apollodorus they were shocking to behold, "long locks dropping from their head and chin and with the scales of dragons for their feet".[13] They pelted Zeus with rocks and he retaliated with thunderbolts.[14] No sooner had Zeus defeated the giants than he was compelled to combat the monstrous Typhon. According to Apollodorus, Typhon hurled flaming rocks from the sky and fire issued from his mouth. Initially the Titans fled from him, but Zeus brought him down with thunderbolts of his own and by throwing entire mountains at him! Ultimately he threw part of Mount Etna at the departing creature, and thus fire still issues from the volcano due to the thunderbolts that were thrown.

The description of Typhon reads like an extreme example of one of the giants.[15] Of human form, his head touched the stars and his hands reached out to east and west. A hundred dragons' heads issued from his hands and coiled snakes from his lower body. His body was described as winged, or feathered. Long hair streamed from his head and fire from his eyes and mouth.

If one prefers to rationalise a myth rather than simply dismissing it then we must ask: what could such monstrous sky creatures be? There is only one possibility. They must remember ancient comets that passed close to the Earth. In the case of Typhon, it seems that Earth even passed through its tail and meteorites fell to the ground. Evidently though, it did not collide with the Earth, for Zeus won this particular battle.

In the version of Apollodorus, the Flood of Deucalion immediately *follows* the war with the giants and the battle in the sky between Zeus and the creature Typhon, but the events are not connected. Therefore, there must have been at least a small time-interval between them. The monster Typhon too, if it remembers a real comet in the sky, must surely have been seen everywhere on Earth.

Indian Cycles of Destruction

Hindu tradition holds that time is cyclical. A kalpa, or day of Brahma is a thousand mahayugas of time. Each mahayuga consists of four ages called yugas. The first yuga, called Krtayuga was a golden age, when people lived long and happy lives; and each successive yuga has seen a diminution in human virtue and quality of life. Our present age is Kaliyuga, which must one-day end in a final confrontation between good and evil.

Another version of the Flood is found in the Vishnu Puranas. The Vaishnavas: Hindus who worship Vishnu, believe that each yuga begins and ends with an avatar, or descent, of the god in the form of an animal. In the oldest tradition, Vishnu was a name of the sun god. The avatar that ended the first age, the Krta-yuga, is associated with the Flood of Manu, when Vishnu returned in the form of a great fish.

Manu was pouring libations by the river, when a tiny fish came into his hands. The fish begged not to be thrown back into the dangerous waters, so Manu kept him in a jar. The fish grew and grew and demanded ever-greater accommodation, until Manu had no choice but to throw the fish in the ocean. Manu recognised the huge fish as Vishnu, who had descended to earth.

The fish (Vishnu) warns Manu of the coming Flood, telling him that in seven days the ocean will rise-up to flood the world. Then a boat will come to fetch him and he must take aboard a representative of every creature. No sooner has Manu loaded two of every plant and animal than the flood begins. Vishnu, who has manifested in the form of a one-horned fish, tows the boat away.[16] All the world is submerged except Manu's vessel and Vishnu, who sits in meditation on the back of the serpent Sesha as he waits for the next world-cycle to begin.[17]

According to the version in the Mahabharata, the end of a world-cycle will be brought about by fire and water. Fire will be seen all around the horizon and seven 'suns' will appear in the sky to dry-up the seas and scorch the land. Then comes the Deluge. Rain will fall continuously for twelve years and the world will be submerged.[18]

Hindu belief therefore recalls *three* floods – a theme that is also evident in Celtic and Chinese lore. The Avatar of the Fish is but one of many cycles. When Vishnu returned in the form of a wild boar, he caused the Avatar of the Boar as he plunged into the ocean. He freed the world from the demons that had usurped it and allowed it to rise out of the sea. A third avatar recalls the churning of the sea when Vishnu reappears, this time in the form of a turtle. Between each avatar, Vishnu sleeps in the waters, lying upon the back of the seven-headed snake Sesha.

Some Indian scholars believe that the story of Manu and the fish may be borrowed from Semitic teachings. A half-man half-fish creature is evident in a very similar story from Babylonian mythology, as recorded by Berossus.[19] He describes a god–like creature called Oannes, with the body of a fish and the head of a man, who appeared at a time when the world was still darkness and water. Oannes taught mankind writing and science, before returning to the sea. Later, but still at a time before the Flood, more of these beings appeared, to continue the education of mankind. The parallels with the avatars of Vishnu are clear and it may be that both myths descend from a common source.

Flood and Deluge Myths from China

The historical period of China is unusually long and this has prevented the decay of history into the kind of aesthetic myths that we find in Europe. As Edward Werner remarked, "…sober, materialistic Confucians and other rationalistic philosophers gradually but surely gained victory over the more fanciful and poetical myth-creators".[20] The Chinese historical epic *Shu Ching* gives us a semi-mythical chronology that commences around 2950 BC. It records the early kings combating devastating river floods during the mid-third millennium BC.

There are two Chinese versions of the Flood. One version is semi-historical in nature. It recalls that in the earliest days, before the Chinese Empire was founded, there was a queen named Nü Wa, whose reign is said to have followed that of her divine brother Fu Hsi (2953-2838 BC). A war took place in which the opposing chief Kung Kung was defeated and she pursued him all the way to the summit of a great mountain. Finding no place to hide, Kung Kung beat his head in anger against the bamboo trees that supported the dome of the sky. The trees came crashing down, bringing with them a part of the sky. Through the tear in the heavens poured a great deluge of water, which submerged the entire world. Only Nü Wa and her followers on the mountainside were saved. Nü Wa is said to have made a paste from stones of five different colours. She repaired the hole in the sky and the floods ceased.

The similarity of the names *Nü Wa* and *Noah* is immediately apparent. Another similarity here is that the waters of the Flood come through a fissure in the sky, or rather the opened windows of heaven as it is termed in the Biblical version.

We may perhaps treat the story of Nü Wa and Kung Kung as a pseudo-historical account of people and events at the time that a flood event occurred. However, other versions of the story are more typically mythical in their style. This is revealed by the description of Nü Wa. Her name means "snail-maid" after her long body and horned head! The Chinese historian Ssu-ma Chien described her as a serpent with a human head – the Chinese dragon. Others say that she had the (horned) head of an ox.

Sometimes the Flood is represented as the result of a battle between the fire god and the god of water. The rebel chief Kung Kung also had some semi-divine aspects to his character. He sought to defeat Nü Wa with the influence of water. In defence, Nü Wa called upon the fire god Mo P'ei. The defeated Kung Kung fled to the west and came to the mountain, about which the sky rotates. In his anger, he shook one of the columns and brought part of the sky crashing down.

The broken pillar *caused the sky to tilt over* and it is recorded that the world "became empty" in the south-east.[21] The waters of the Deluge flowed in through the breach in the firmament and the world was devastated before Nü Wa could set things right again. She is said to have

Figure 1.1 Multiple suns appeared in the sky. According to Chinese myth, ten strange 'suns' appeared in the sky during the reign of the patriarch Yao. (Reproduced by permission of Penguin Books)

repaired the hole, first with the ashes of reeds and then by melting stones of five different colours. To stop the ground shaking, Nü Wa cut of the feet of the celestial tortoise, upon whose back the world rests.

It may therefore be seen that the Flood is deeply embedded in Chinese cosmogony, for it also explains the tilt of the Earth's axis. Ever since Nü Wa repaired the sky, it has shifted to the north-west and the stars have changed their courses. For this reason, the mountain that supports the dome of the sky was known as "the imperfect mountain".

The Shu Ching records two more flood episodes in China. Devastating river floods are described during the reign of the patriarch Yao (2357–2355 BC) and again around 1766 BC (according to the legendary chronology) heaven and earth conspired against the evil emperor Chieh.[22] Two suns fought in the sky and the mountains crumbled. There is no archaeological confirmation of any of this and later historians refused to trust the legendary dates. After Chieh, the historical Shang Dynasty commences. This too was considered legendary, until the discovery of the oracle bone divinations during the 1930's. It is reasonable to conclude that Chinese chronology will not allow us to date the Flood more recently than the early third millennium BC.

Myths from Northern Europe

In the Icelandic poem *Voluspa*, or 'Sybil's prophesy' dating from about AD 1000 we find a North Germanic myth of the death and rebirth of the world. It reads as a prophecy, as if the current cycle would repeat the previous one, and so on forever.[23]

The Norse view of the universe was one of three levels: heaven, earth and underworld. The home of the gods was Asgard where the warrior-gods dwelt in the great hall of Valhalla. Nearby lay the rectangular plain of Vigred; described as a hundred miles (or a hundred and twenty leagues) square. Beneath Asgard lay Midgard, 'the world of men', surrounded by the vastness of the world-ocean. In this ocean lived the world serpent, Jormungand; so huge that he circled the world and could bite his own tail. Below lay Nifflheim, or Hel, the land of the dead.

In the Norse myths, the Flood is brought on by the battle between the gods and the Midgard Serpent. The fall of the gods, Ragnarok, will begin with endless wars and people will commit forbidden sins. Next comes the experience of a great winter, the Fimbulvetr, when three terrible years pass without a summer and the snow persists on the ground.

The wolf Hati will catch-up with the Moon (an eclipse) and the stars will vanish. Monsters are let loose and the World Serpent rises out of the ocean as waves beat against the shore and the land shakes. The serpent Jormungand and the wolf Fenrir will do battle in the sky. The serpent will spew venom upon the earth and the sky itself will be torn asunder.

The gods flee from Asgard. The giants and the gods meet one final time on the great plain of Vigred. There they do battle as the world slowly sinks into the sea. Yet, the world will be reborn from the ice and cold just as it was in the beginning. A new green Asgard emerges out of the sea; and the gods build a great wall around it to keep out the giants.

In the Welsh Triads, we may find another European memory of the Flood, which could predate any Christian influence. One of the *three awful events* is the bursting of the lake of waters, which drowned all mankind except for Dwyvan and Dwyvach, who escaped in a boat without sails. The second calamity was a storm of fire, which split the Earth apart and consumed most of the people. The third calamity was a scorching summer that burned man, beast and trees alike, so that some species were lost forever.[24]

In another triad, the survivor is Nevydd who escaped in a ship along with a male and female of every creature, "when the lake of waters burst forth".[25] Nevydd is clearly Noah and Dwyvan, or Dylan, seems to be the Greek Deucalion. However, the lake of waters is here an imagery of the ocean and it may be seen that the flood is an overflow of the sea rather than a deluge from the sky.

It has been argued that these triads are recent forgeries, based on Biblical and classical myths, but the association of the flood with a lake burst would argue that the memory is ancient. A deluge of rain is nowhere mentioned, and lake-bursts are common elsewhere in Celtic myths and legends. The calamities of fire and heat resemble those of eastern mythology, which were surely unknown to the medieval Welsh.[26] That the Atlantic Celts indeed had such a myth is further testified by the Greek geographer Strabo, who records their belief that at the end of the world, "fire and water would ultimately prevail".[27]

In other Celtic legends, we may find memories of a flood that submerged the coasts around Britain in ancient times. An eleventh century chronicle records another Welsh triad that tells of the submergence of three ancient kingdoms: the lost lands of the kings: Helig, Teithi Hen and Rhedfoe;[28] and Helig's palace is said to lie submerged in Caernarfon

Bay. These may be the same as Cantrae'r Gwaelod or The Lowland Hundred, likewise said to be submerged in Cardigan Bay. Another lost land called Lyonesse is popularly believed to have lain between Cornwall and the Isles of Scilly; and similarly the city of Ker-Is off Brittany. These submergences are not embodied in the mythology, but are legends, which appear to lie just before historical record.

Flood Myths from the Americas

Among the Aztec beliefs that the Spanish found in Mexico was the story of the five suns. The Aztecs believed that their gods had created and destroyed four previous worlds; and that we now live in the fifth cycle of creation. Like their Maya predecessors in Central America, the Aztecs were obsessed with time and the calendar. They believed that unless they placated their gods with timely human sacrifices and rituals, then the world would be destroyed again. The sun god had to be nourished every day with blood from a beating human heart, or he might not retain enough strength to rise again next day.

In each Aztec cycle of creation, the sun was reborn with a different name. Each sun was sponsored by a different deity, who created a race of people. The first world was inhabited by giants but they were eaten by (presumably giant) jaguars. The second sun was destroyed by winds and the surviving people degenerated into monkeys. The third creation was consumed in fire and muddy rain and its people became dogs, butterflies and turkeys! The fourth creation was ruled by the goddess of water and was terminated by a flood; its survivors being transformed into fish.

Our present sun is named Tezcatlipoca and was said to be the work of Quetzalcoatl who, together with the sun god, became world-trees and raised the sky aloft. First however, they had to battle a huge alligator creature that tried to prevent the creation. This monster is sometimes represented as an aspect of the earth god. Quetzalcoatl then retrieved from the underworld the bones of the people who had died in the flood and the two gods gave some of their own blood to create the first flesh of humanity. The gods then sacrificed themselves on the pyre at Teotihuacan and one of them was transformed into the new sun.

Most people will have heard of the Aztec god Quetzalcoatl, whose name literally means 'feathered serpent'. He was believed to have been expelled by a more powerful god and sailed away into the Atlantic, promising to return. One description says he had white skin, long hair and a flowing beard – a description that is reminiscent of the Greek Typhon. When the bearded Spaniards arrived in the sixteenth century, the Aztecs believed that their god had returned and so they sent a gift of the snake mask and feathered cloak – the emblems of Quetzalcoatl.

The gods of the older Maya civilisation are known only from their surviving inscriptions and documents. A fragmentary record of the myths and traditions of the Quiché Maya is contained in the Popol Vuh, a document probably dating from around the time of the conquest.[29] Their cosmogony myths show many similarities to those of the Aztecs.

In the Maya version, the multiple creation is explained by a degree of ineptitude on the part of the gods. The twin creators Gugumatz and Hurucan wish to make a race of people who can worship and nourish them. When first they try to build people out of mud, they are too soft and rapidly dissolve. They condemn the first race to be the animals and to be forever food for higher creatures.

They next attempt to make men from wood and women out of reeds and rushes, but although they can speak, they are too simple-minded to worship the gods in the required manner. Therefore, the gods decide to destroy their creations in a great flood. They also send destruction from the sky and turn the people's own dogs and household artefacts against them. Even so, a few people escaped and they survive as monkeys.

The Popol Vuh then tells a myth of the Hero Twins Hunahpu and Xpalanque, demi-gods, who battle with the celestial macaw Vucub Caquix. With the aid of his sons, the macaw has set himself up as a *false-sun* in the sky. They deprive him of the jewelled teeth that make him shine and render him unable to eat. Thus, the false sun and his offspring are soon vanquished.

Next, the gods try to make people out of maize flour. This time, they experiment with the first four humans. The gods find that their creations are too perfect and worldly-wise; and so they place a limit to their vision, so that henceforth mankind may never fathom the purpose of the gods. We are told that the four founders of humanity journey in darkness to the Mountain of the Seven Caves. There, the god Tohil promises mankind the secret of fire, but only after they perform human sacrifice to him. Only then do the gods allow the first true sun to rise in the sky.

We therefore see once again that mythology is religion. These are not just naïve stories. The very function of humanity is to nourish the gods, just as the animals are here to nourish us. Priests would commit daily ritual murder to feed their gods with blood, lest they should again decide to destroy the world. Ultimately, they knew, the world cycle would end at the appointed time.

In the myths of Central America, we find all the elements that we have seen elsewhere. In addition to the flood, there is a memory of a battle in the sky between the gods. Our present sun is not the same as the sun before the flood. We see multiple sources of radiance in the sky. The imagery may differ from Old World mythology, but the underlying concepts are the same.

A World-wide Deluge?

Flood myths are found in the folklore of peoples as widely separated as Eskimos and Australian Aborigines; from South American Indians to Polynesia. There are simply too many to describe all of them here.

In the Australian Aborigine version it is a great snake or a fish that crosses the sky and drags the flood waters behind it.[30] The Greenland Eskimos believed that the world had once tilted over and the sea flooded the land. Only one man survived, but when he struck the ground with his stick a woman emerged; thus was the world repopulated.[31] A Peruvian version of the flood recalls that only one man and a woman survived the flood floating in a box. When the waters subsided they were far from home, and so they fashioned people from clay and dressed them in strange clothing. When these came to life they tunnelled into the ground, to emerge in various parts of the world as the different races of mankind.[32]

An Amazonian myth described by Levi-Strauss has a novel origin for the waters of the flood. The brother of the creator-god gave humanity the fruit of the tree of life, but they discovered where he obtained the food and began to steal it for themselves. Angry at such misbehaviour, the creator-god cut down the tree of life and the waters of the flood issued from the broken stump.[33] This is typical of many tribal myths, which seek to explain the origin of the floodwaters by a naïve transference of the problem.

However, some areas of the world do not appear to have a flood myth at all and it may be that not every part of the world was affected in the same way. Flood stories are difficult to find in the myths of the African continent and some researchers have even declared them non-existent. The ancient Egyptians do not seem to have preserved a detailed Flood myth – although this is perhaps to be expected in a land where flooding was an annual event. However, Plutarch does tell us that a deluge of rain was associated with the Myth of Osiris.[34] Flood myths are also difficult to find among the mythologies of Japan and Oceania; and the same is true of the Siberian and Finnic races. However, there is no shortage of cataclysmic events of other kinds. The Japanese myths prefer to describe the creation of the islands out of the sea, possibly associated with the earthquakes and vulcanism that prevail there.

Oceanic myths revolve around the enigmatic figure of Maui, the demigod who fished the islands out of the ocean in his net. New Zealand Maoris therefore call their island The Fish of Maui.

Maui was said to have been born only four generations after the creation of the world, while the process remained incomplete. A myth from Tahiti says that he had eight heads! Some island myths give him responsibility for raising the sky aloft; others say that he arranged the stars and snared the Sun in a net to make it move more slowly. Evidently then, Maui was another sky-god, like Tiamat or Quetzalcoatl, but here we find him associated with benevolent, rather than destructive forces.

Thus, we see attached to Maui, many of the feats that elsewhere are associated with flood and cyclical creation myths; but this time, rather with the rise of land out of the sea; the opposite of the flood.

Serpent or Dragon?

The Hausa people of Nigeria have a serpent myth that is very similar to that from Scandinavia. They consider that all water comes from beneath the earth. After the creation the weight of mountains and trees became too great for the world to carry. The creator therefore fashioned a great serpent, which coiled itself around the earth to hold it upright and so prevent the world from sinking into the sea. When the snake wriggles, he causes an earthquake. Yet, he does not emerge to cause a flood.

The theme of a *sky creature* as the cause of the flood is surprisingly common; and as we have seen, it is also prominent in areas where the myths recall *rising* rather than *sinking* of the land. Sometimes the agent is a great fish, or a crocodile, but more usually a serpent or a dragon. The world-wide occurrence of these serpent myths would argue against those who hold that flood myths, in the Americas and elsewhere, result from recent contamination by Christian missionary teaching. Missionaries also reached Africa, where there are few, if any flood myths – yet the serpent myths are strong.

Some aspects of the Chinese Nü Wa or Kung Kung suggest a horned serpent or dragon. The Greek Typhon, as we have seen, was serpent-like and perhaps the same may be said of the shapeless Babylonian sky-god Tiamat, who is sometimes likened to a dragon. In various Biblical references, we glimpse a memory of a battle between Yahweh and a great monster during the creation of the world. He is variously Leviathan, or Rahab, the fleeing serpent.[35]

The serpent is also associated with cycles of creation. The Aztec Quetzalcoatl was a feathered serpent and could therefore fly. Meso-American myths recall an alligator beast that tries to prevent the creation. The World Serpent of Norse mythology emerges at the end of each world cycle. Similarly, the Indian beliefs recall the emergence of dragon-like creatures at each avatar, when the world is destroyed and reborn.

Astronomers Victor Clube and Bill Napier have argued convincingly that these various serpent myths recall aspects of ancient comets that passed close to the Earth.[36] Their differing shapes and aspects gave rise to animal symbolism and their occasional forward-pointing jets were likened to the horns of a beast.

Battles in the Sky

Greek mythology offers us a memory of a battle in the sky. This is found in the war between the Titans and the giants; and between Zeus and the monster Typhon. The Norse myths similarly have the wolf-god battling with the serpent Jormungand at the end of a world-cycle. The Chinese flood myths apparently reveal a more conventional battle between gods, but its outcome affects the sky and alters the very rotation of the stars.

The Babylonian equivalent of this battle between the gods and a sky dragon is found in that between Marduk and Tiamat. In the creation epic *Enuma Elish*, Tiamat is the ocean goddess and Apsu personifies the 'sweet water' beneath the earth. Tiamat is often represented by a dragon, but the description of her is vague and shapeless. She coils like a snake but retains a moisture; and she has a shell more like some vast mollusc-creature, akin to the Chinese Nü Wa.[37] Lahmu says of her:

> What a strange and terrible decision, the coil of Tiamat is too deep for us to fathom.

From the union of Apsu and Tiamat the whole world is born out of the waters. They first give birth to two serpent creatures and then to the dynasty of the gods in turn, who spread throughout the world. Soon however, Apsu is irritated by the noise of all these new gods and plots with Tiamat to destroy them again. Ea overhears the plot; he warns the other gods and sends his son Marduk to do battle with Tiamat.

Tiamat gives birth to an army of huge serpents and to eleven other fearsome creatures, among them, a worm and a dragon.[38] Marduk challenges Tiamat to battle and ensnares her in a great net. He cleaves the body of Tiamat in two and uses it to fashion the dome of the sky and the basin of the ocean. Next he sets-up each of the gods as a constellation and creates the calendar:

> ...he measured the year, gave it a beginning and an end, and to
> each month of the twelve, three rising stars.[39]

Thus, Marduk emerged as the highest of the Babylonian gods. He created Nebiru the pole about which the universe rotates.[40] He assigned each of the gods to a proper place as stars in the sky; we are also told that he set the length of the year and regulated the motions of the planets. Evidently then, before these mythological events, the year and the motions of the planets were not as we know them today.

The parallels with the Greek myths are apparent: Marduk is Zeus and Tiamat is Typhon and the army of serpents that she creates are the gigantes who battled with the Titans. We are given the additional detail that the battle determined many astronomical phenomena and the calendar. However, whether the episode is the same catastrophe as the Deluge is not at all clear. The cause of the battle between the gods is the noise and clamour of the minor deities, just as the din of humanity is the excuse for sending the Flood.

Falling Stars

In the Judeo-Christian tradition the cause of the Flood has been lost in the various revisions of the Bible. However, in the apocryphal Book of Enoch, dating from the early centuries BC, we may glimpse a view of the antediluvian world that has slipped out of the Biblical texts.[41] Like the Babylonian Gilgamesh, Noah's ancestor Enoch makes a visit to Sheol, the underworld: a place in the west, where fallen angels are imprisoned. There, in a dream he foresees the coming Flood.

In Enoch's vision, the flood will be a punishment of the angels who have been banished from heaven. That these angels are *fallen stars* is evident in many contexts:

...These are of the number of stars of heaven which have transgressed the commandment of the Lord, and are bound here till ten thousand years; the time entailed by their sins, are consummated.[42]

And again:

...I saw in a vision how the heaven collapsed and was borne off and fell to the earth. And when it fell to earth was swallowed up in a great abyss.[43]

Here again we can clearly see that the Deluge has an *astronomical* cause; and it not only devastates the antediluvian world, but also affects the sky and the movements of the stars and planets:

And many chiefs of the stars shall transgress the order prescribed:
And these shall alter their orbits and tasks,
And not appear at the seasons prescribed to them.[44]

It must be remembered however, that the apocrypha and other apocalyptic literature record an oral tradition written down in their extant forms as much as two thousand years later than their Egyptian and Babylonian equivalents. The Jewish historian Josephus considered that the actions of these fallen angels "resembled the acts of those whom the Greeks call giants".[45] Thus, he provides for us the missing link between the battle in the sky and the deluge that follows it – the Flood had an astronomical cause.

One may find some surprisingly similar wording in British mythology, which was presumably preserved by the Welsh Bards from the oral traditions of the British Druids. This has come down to us in the form of the Prophecies of Merlin. These too, tell of a time when the stars (or rather the planets) altered their courses.

The stars will avert their gaze from...men and alter their accustomed course. The harvests will dry up through the stars' anger and all moisture from the sky will cease...

The malice of the planet Saturn will pour down like rain, killing mortal men as though with a curved sickle. The twelve mansions of the stars will weep to see their inmates transgress so...

The Moon's chariot shall run amok in the zodiac and the Pleiades will burst into tears. None of these will return to the duty expected of it...

In the twinkling of an eye the seas shall rise up and the arena of the winds shall be opened once again. The winds shall do battle together with a blast of ill-omen, making their din reverberate from one constellation to another.[46]

We could not hope to find a clearer description of a flood brought about by an astronomical agency.[47] The Earth wobbles on its axis, the sea spills over the land and subsequently the Sun, Moon and planets no longer conform to the established calendar. The prophecies also recall an era when the channel between Britain and France receded to a mere trickle and a man on one shore could shout to a man on the other.[48] Not only does the sea flood the land, but it may also recede to create rising land. Who would ever dream of such things if they had not actually seen them? And it is all here, in the much-neglected British Celtic mythology.

Lost Islands of the West

Yet other myths suggest that rather than being universal, the recollections of an ancient flood could perhaps be *localised* to a memory of an ancient cataclysm along the Atlantic coast. In the pages of Diodorus of Sicily, we find a description of a wonderful land that he believed lay far out in the Atlantic Ocean. I can think of no better tribute to both this much-maligned historian and to his translator, than to quote the passage.[49]

...For there lies out in the deep off Libya an island of considerable size, and situated as it is in the ocean it is distant from Libya a voyage of a number of days to the west. Its land is fruitful, much of it being mountainous and not a little being a level plain of surpassing beauty. Through it flow navigable rivers which are used for irrigation, and the island contains many parks planted with trees of every variety and gardens in great multitudes which are traversed by streams of sweet water; on it also are private villas of costly construction, and throughout the gardens banqueting houses have been constructed in a setting of flowers, and in them the inhabitants pass their time during the summer season, since the land supplies in abundance everything which contributes to enjoyment and luxury. The mountainous part of the island is covered with dense thickets of great extent and with fruit trees of every variety, and inviting men to life among the mountains, it has cosy glens and springs in great number...And speaking generally, the climate of this island is so altogether mild that it produces in abundance the fruits of the trees and the other seasonal fruits for the larger part of the year, so that it would appear that the island, because of its exceptional felicity, were a dwelling-place of a race of gods and not of men.

Diodorus tells us that in his times this island remained undiscovered. The Phoenicians from the earliest times made voyages of trade and established colonies in North Africa and Iberia. They extended their exploration into the Atlantic and as they voyaged further out, storms drove the Phoenician sailors ashore on the fabled island.

In the words of C. H. Oldfather, "the idyllic colours in which the picture of this island in the Atlantic is painted relieve the historian of any concern over its identification"; and this comment is illustrative of an attitude adopted by many scholars of his generation when confronted with mythology. In other words: to accord something the status of myth or legend is to reduce it to mere fiction and then its content can be safely trivialised and ignored.

Diodorus Siculus was a Greek historian who lived in Sicily between about 80 BC and 20 BC. His Library of History was an attempt to catalogue the history of the known world, from the earliest times up to Caesar's conquests. The purists will say that he had a tendency to *euhemerise*, that is: to rationalise mythical figures after the style of the historian Euhemerus (c 300 BC), as rather a memory of real kings and heroes who were deified after death. Other Greek historians had tended to neglect the traditions, leaving them to be related by the specialist mythographers. However, in treating myths as history, Diodorus has preserved details not found in more respected sources such as Ovid and Apollodorus.

It is ironic that Diodorus continued his own survey of the islands in the Atlantic Ocean with a report of contemporary Britain, only recently advertised to the Roman world by the campaigns of Caesar. However his description drew mainly upon Pytheas, who charted the Atlantic coasts at some time around 300 BC. Diodorus does not equate the two islands, and there is really no reason why he should have done so – for the two descriptions must be separated in time by more than a thousand years. We may therefore forgive Diodorus for believing that some other large island lay opposite Spain, but modern scholars should surely know better. When the earliest classical geographers report islands in the Atlantic that can be positively identified with Britain and Ireland, they are often described as lying 'opposite' Iberia;[50] and this is quite true: they do lie directly opposite the north coast of Spain. The 'island of considerable size', *if indeed it were a real place,* can be none other than a very ancient description of the British Isles. Rather than properly investigate this possibility, modern scholarship has conveniently left it hanging as a myth.

In his third book, Diodorus relates a further myth about an island, evidently drawn from a quite different source.[51] He called it simply Hespera, from its position in the west. Indirectly, we may deduce that this island too was in the Atlantic. It was home to the warlike Amazon women and nearby lived a people called the Atlantians, whom they conquered. These too he places by the coast of the ocean – although he does not locate them on the island itself. Hespera too was an island of great size, full of fruit trees and domesticated herds. We may recognise this as another, quite distinct, memory of that same island that the Phoenicians believed they had found.

We then see why later explorers were unable to recognise the fabled island. Diodorus records that many regions along the Atlantic coast were 'torn asunder' by a catastrophic earthquake and disappeared. So it seems we have a number of distinct mythological sources, two at least, which recorded the existence of a once flourishing island civilisation. One stream of thought believed the island still extant in classical times; the other held that the island and its deified inhabitants had disappeared long ago during a widespread catastrophe that devastated the Atlantic coasts.

In the tradition of Diodorus, his Atlantians were none other than the Titans and heroes of Greek mythology, as he says:

> But since we have made mention of the Atlantians, we believe that it will not be inappropriate in this place to recount what their myths relate about the genesis of the gods, in view of the fact that it does not differ greatly from the myths of the Greeks…and the gods they say were born among them.[52]

He goes on to relate familiar myths about Uranus, Cronus, Atlas and the other Titans of the Greek pantheon, making them the Atlantian royal family. We may see a comparable rationalisation in the works of the Jewish philosopher Philo Judaeus. He interpreted Greek myths as derivatives of the Phoenician; equating, for example, the Phoenician god Ouranos with Uranus and his son El with Cronus. El is said to have lived in a place called the Sad El or Fields of God, which are equivalent to the Elysian Fields of the Greeks and Egyptians.

Apollodorus, who is probably our most respected source for Greek mythology, makes no mention of any Atlantians. However, even he records that Heracles was sent to a northern people known as the Hyperboreans to seek the golden apples of the Hesperides;[53] and it was there that he encountered the Titan Atlas, after whom of course the ocean takes its name. This at once raises the question whether Atlantians and Hyperboreans should be regarded as one and the same.

Atlantis – an Ancient Egyptian Flood Myth?

The most controversial account of the catastrophic submergence of an island in the Atlantic Ocean is that given by Plato; a story inherited from his ancestor Solon, who had in turn acquired it from the priests of Saïs in the Egyptian Delta.

Solon visited Egypt around 590 BC and engaged in discussions about the Flood of Deucalion with two priests of the goddess Neit. They derided him, saying that there had been many great floods in the past but the childlike Greeks had forgotten them all. The priest then related the story of Atlantis.

In Plato's *Timaeus* and *Critias* the lost island is said to have lain opposite Gadeira and the Pillars, and the Titan Atlas was its first king. The significant difference is that Plato offers the destruction of Atlantis, not as mere mythology, but as a true history complete with a chronology. We see the same idyllic setting that we encountered above in the fabled island of the Phoenicians. Atlantis too was a large island with mountains and springs; and with a mild and genial climate. The island possessed a rectangular central plain, transgressed by navigable rivers and canals, which converged upon a great city. People inhabited both the plains and the hills above. They planted various fruit trees and grew wealthy on the abundance of forests and game. According to Plato, the island sank into the ocean in the course of a single day and night of catastrophic floods, together with earthquakes that were felt even as far away as Greece.

Classical scholars have tended to dismiss the idea that Solon could have brought the story from Egypt as Plato claimed, on the grounds that no such island, nor indeed any strong flood myth, is known from any Egyptian inscription. Some have dismissed the story as a fable invented by Solon or Plato, based upon memories of Crete and the Aegean world.

However, the investigator will find more similarities between Plato and Diodorus than with any Greek myths. Other than Atlas, Plato never mentions the Titans; and there are other subtle differences, which would indicate that the two writers drew their information from quite independent sources. If Diodorus had known of Plato's Atlantis then surely he would have linked it with his own Atlantic island, or with his own Atlantians? Yet, as we have seen, he does not even acknowledge a similarity between his own two accounts of the islands in the west.

The Antediluvian World

If we set aside for the present these indicators of an ancient cataclysm, we may see within the idyllic setting of these islands another view of the world as it was *before* the destruction; a view of the *antediluvian* world, if one may utilise that somewhat outmoded term. We should ask: could these be genuine memories of a former era, when the climate and coastlines of the world were not as we know them today?

The Biblical view of the antediluvian world, which again is probably modelled on Mesopotamian beliefs, reveals a belief in a former, gentler age: a paradise. The creation myth of Adam and Eve in the Garden of Eden reveals a belief that there was formerly a better time for mankind; and that as a penalty for our own sins we are obliged to toil for our living. The same idea too, is evident in classical descriptions of a golden age.

Indian beliefs too, recall a golden age before the Flood, although we cannot be sure whether this cosmogony has a similar origin to the Babylonian and western myths. In Egypt too, we see that the memories of the former world may have lingered on in their view of the happy land of the dead: The Elysian Fields.

The Paradise of the Dead

The Egyptian kings of the pyramid age were entombed along with elaborate spells and instructions to guide them in the afterlife. By the later New Kingdom many quite ordinary persons were also interred with papyri depicting these spells. Some papyri show maps of a place called *Seket Hetep*, the 'Field of Offerings' or 'Field of Reeds'. These depict a plain with a rectangular grid of canals, variously cut through by rivers; and with a city where the gods were believed to reside. Sometimes this place was located to the west of Egypt, towards the setting sun.

Once stripped of its religious context, the Egyptian paradise of the dead is not unlike the pleasant land described in the pages of Diodorus Siculus, or the rectangular plain described by Plato; and it is not out of the question that Solon was told something of this mythology by his Egyptian tour guides. For example, inscribed in Pyramid texts we find:

> These are they whom Nut bore…who make the Nt-crowns of the canals of the Field of Offerings for Isis the Great…I have gone to the great island in the midst of the Field of Offerings on which the swallow gods alight…[54]

The elevation of this paradise to mythology, as a sacred land of the dead, must therefore be dated *no later* than 2500 BC. This shows us that any real place upon which it was modelled, and the flood-cataclysm that destroyed it, must be generations older still.

The Ptolemaic kings introduced Aegean influence and much of Egyptian religion came to be equated with that of Greece. The Greeks recognised the Egyptian land of the dead as the same place as their own Elysian Fields.[55] We see these in Homer's *Odyssey*, where Menelaüs on his deathbed is promised that he will go to the Elysian Fields, at the world's end, where the westerly breezes blow.[56]

Other mythological contexts reveal to us that this land of the dead was the same place as the Island of the Blessed in the western ocean. Hesiod describes how Zeus revives (in afterlife) the heroes of Troy:

> The son of Cronus gave the others life
> And homes apart from mortals, at Earth's edge
> And there they live a carefree life, beside
> the whirling Ocean, on the Blessed Isles
> …and Cronus is their king, far from the gods.[57]

Here we glimpse the classical poetic view of this idyllic place, which has shaped the view of generations of classical scholarship. By the age of Homer and Hesiod, around 700 BC or thereabouts, any historical context for Cronus and the other Titans had long been lost. Later mythographers such as Pausanias would consider Elysium as merely a division of Hades, the land of the virtuous dead.

Such concepts are not confined to classical sources. Other European races preserved similar myths in their oral traditions. We find more evidence in the Irish myths of a mysterious *Otherworld*, which may be glimpsed in a number of legendary tales. Most such stories must be placed in a category of fiction that has evolved around real heroes and settings, completely lost in time. The Otherworld is again a beautiful land, with plains and rivers, where a golden-haired race live in high towers and castles. There is never a sign of winter; it is *The Land of Summer*. At other times, it is *The Plain of Delights*; and another of its names is *The Land under the Wave*, revealing it as a plain that sank beneath the sea. It was watched over by Manannan mac Lir, the Celtic god of the sea, after whom the Isle of Man is named.[58]

The myths of the ancient Finns are preserved in the epic songs of the *Kalevala* (The Land of Heroes). These were not recorded in writing before the early nineteenth century. Here too we find a paradise of the dead called Manala, or alternatively Tuonela, the home of Tuoni, who ferries the dead across the underworld river. That Manala was perceived as an *island* is apparent in the description of Väinämöinen's journey:

> Then to Tuonela he journeyed...
> Straight to Manala's dread island
> And the gleaming hills of Tuoni[59]

Just like Odysseus and Gilgamesh, the hero Väinämöinen was able to visit this place while still living, for Tuoni's daughter asks him:

> Hence a boat shall come to fetch you,
> When you shall explain the reason
> Why to Manala you travel
> Though disease has not subdued you
> Nor has death thus overcome you[60]

Some have attempted to derive even the ideas of Finnish mythology from classical influence, but such diffusionist thinking is surely outmoded;

and its proponents must explain why the Finnic myths more closely resemble the Egyptian than the Greek.

Another memory of the land of the dead in the guise of an idyllic island may be seen in the misty Isle of Avalon, where King Arthur, like Cronus, is taken at his death and waits in eternal sleep until again called upon. We may find a familiar description of Avalon in Geoffrey of Monmouth's *Life of Merlin*:

> The island of Apples, which men call the Fortunate Isle, is so named because it produces all things of itself. The fields have no need of farmers to plough them, and nature provides all cultivation. Grain and grapes are produced without tending, and apple trees grow in the woods...[61]

Surely then, it is no heresy to suggest that the classical myths of an island-paradise in the Atlantic, destroyed by a great flood, were drawn from Carthaginian and other North African histories. Indeed, they may have preserved them in a much better state than their classical equivalents in the Graeco-Roman world.

Extremes of Weather

The transition of the world from the idyllic climate of Hesiod's Golden Age to its present miseries is echoed in a variety of other mythical sources. Some of these we have already discussed.

The Roman poet Ovid, while exiled from Rome, composed a poetic history of the world, known as the *Metamorphoses*. We cannot be sure what sources he used, but he contributes much that is familiar as well as much that is unique. He tells us *of a time when the climate changed*; when the age of eternal springtime ended and our four seasons began. The transition was one of weather extremes, when the world alternately sweltered or was icy and frozen. He dates this event to the time when the silver race replaced the men of the golden age, and that this was in the reign of Jove (Zeus) when Saturn (Cronus) fell to the underworld.

In the Egyptian Myth of Osiris, the emergence of the Nile Delta is again linked to the time of the Deluge. In the Northern myths, we also see that the flood which ends the world cycle is accompanied by intense cold and snow, rather than rain. In Celtic mythology, more specifically the Welsh Triads, we are told of extremes of heat that killed many plants and animals. Are these to be seen as memories of the same event, or of many similar occurrences; or shall we simply ignore all this valuable evidence about the past?

History or Symbolism?

It has been the fashion since the work of psychoanalysts such as Carl Jung and the anthropologist Lévi-Straus to view myths as a kind of primitive symbolism, which arise naturally out of human dreams, to explain our origins and events beyond the primitive comprehension. It will be apparent to the reader that the present author has little patience with this view, which has contributed to a neglect of the evidence contained in myths. Even if some myths do arise by such a process then it still does not explain where all the history has gone.

Everyone surely, has heard the joke about the message passed orally along the British trenches of the First World War. "Send reinforcements, we're going to advance" is transmuted into, "send three-and-fourpence we're going to a dance"! The truth lies at the beginning of the chain, not at the end and however much we analyse the message in its final form it will never tell us what the general really intended.

Even in ancient times, people told their children about their heroic ancestors, but true history is preserved only when recorded in a form that retains the sequence of events and allows them to be assigned to an era. Many ancient peoples, as will be discussed below, had no reliable calendar era. It is only in literate societies like China and Egypt that calendars and records have delayed the loss of historical sequence.

We may therefore consider a tradition, or a legend, as an historical event that has become detached from its chronology, such that we no longer know for sure when it happened, or even if it is true. It becomes, so to speak, a 'particle', free to wander.

We tend further to make the distinction that a legend should be termed a myth when it is associated with the acts of deities or demigods (i.e. persons deified after death) or with cosmogony beliefs – but there is really no clear division. A further unfortunate connotation is that a myth is also used as a euphemism for something fictional.

Several things may happen when a historical particle becomes detached from its place. Firstly, of course, there may be a loss of important detail. Secondly, events may undergo a kind of 'Brownian Motion' as the particles become jumbled out of sequence in the retelling.

Another consequence is the tendency to aggregate characters of similar name or nature into a single personality; or to aggregate similar-sounding events into a single event. Details may also transfer between similar events or persons. Consider a hypothetical example from British history. Suppose that our descendants five thousand years from now encounter a mythical tyrant named 'Cromwell'. He was said to have usurped power, beheaded the king and reformed the religion; yet became so inept in government that he was caricatured 'Tumbledown Dick' and himself later executed for his treason. Because we have the history, you may recognise a composite of Thomas Cromwell: the Tudor Lord Chancellor, with Oliver Cromwell, plus a little of his son Richard thrown in. We are fortunate, that in some legends we can actually see this process of composite character formation beginning to occur.

Again, as the events recede into the past, only the strongest characters and the most momentous incidents will survive and it is inevitable that they will become embodied into religious beliefs. All religion commences from a fundamental human need to explain birth and death; and from the reverence of the ancestors. It is inevitable that the ancestors of kings and tribal chiefs should become deified, in the same way that Roman emperors were elevated after death; and the Egyptian pharaohs even during their lifetime. Tyrants have always sought to control their populations by claiming a special relationship with the gods. The population may fail to obey a mere king, but they will surely fear a god! One man's cosmogony myth is another's religion and once history falls into the hands of priests, it becomes diverted to an altogether different purpose.

Fictional storytelling may also preserve the true context in which its characters move. In written form this is conserved, but in an oral tradition, this background is continually modernised. Oral history may even survive a change of language. The names of people and places may change and lose their original meanings. Real hero figures may be made to follow a fictitious plot. We see this, for example, in the medieval revisions of the legends surrounding King Arthur, already a composite figure of earlier legend; again in the tales of Robin Hood; or the heroes of the American Wild West. Sometimes, we may find relict details that allow us to suggest a date for the origin of a particular story.

Therefore, when we examine myths and legends we must expect that they may have undergone any or all of these processes. If we view them as *decayed history* rather than as symbolism then it may be possible to recover from them some truth about the ancient past. Instead of dismissing myths and legends, we should analyse their content with the same scientific rigour as any physical artefact uncovered by the archaeologists. This is particularly important when our goal is evidence of *catastrophic* or *astronomical* events in human prehistory. As these events were experienced world-wide, the same memories may be preserved quite independently in the myths of many different races, allowing us to extract and compare the details.

It is worthwhile then, to summarise what can be learned from the preceding analysis, which can itself be a mere summary extracted from a myriad of myths and legends.

Table 1 summarises the occurrence of traditions of an *earthly paradise* and its *destruction*. We may recognise a memory of an earlier epoch when *the climate was milder*, and this is frequently associated with an *idyllic island* or a *paradise*, sometimes located in the west. The 'gods' lived in this paradise, sometimes in a city, and there too go the deified heroes after death. We may then see an astronomical or astrological event, symbolised by a battle with a *sky dragon* or *serpent* creature, which terminates this golden age. This time may be associated with fire or *extreme weather* and deterioration of the climate. Sometimes a *flood* or *deluge*, or an *overflow of the sea* is associated with the event. Not all mythologies conserve every aspect, but the degree of overlap enables a composite picture to be reconstructed.

Sometimes we can assign an upper or lower limit to the date when the events in a myth must have occurred. Where we see similar events described by quite unrelated cultures then we may infer that they predate the *oldest* culture for which we hold reliable history. Events that are 'mythical' in one culture, such as the Celts or the Greeks, may overlap the period in another culture for which written history survives: principally Egypt, Mesopotamia and China. Moreover, once we cease to ignore the message of the myths then we have an opportunity to compare the references to events such as floods, weather and famine, with directly dateable physical evidence.

The degree of authenticity that we may assign to each mythology is conferred by how old its written sources are. The oral traditions of the Basques and Finns were not written down before the nineteenth century. Celtic myths are generally found in early medieval documents; and have suffered much distortion of the pagan beliefs during Christian conversion.

Table 1	Flood or Deluge	Sky Dragon/ Serpent	Idyllic Climate or Changes	Earthly Paradise/ Golden Age	Idyllic Land of the Dead	Sunken Land or Islands	Emerging Land or Islands	Atlantic Islands
Graeco-Roman	✓	✓	✓	✓	✓			?
Phoenician	✓				✓			?
Germanic	✓	✓				✓		
Welsh	✓	✓	✓	✓	✓	?		✓
Irish	✓	✓	✓	✓	✓	✓		✓
Mesopotamian	✓	✓	?	✓	✓		?	
Chinese	✓	✓	?	✓	✓	?		
Indian	✓	✓	✓	✓			✓	
Meso-American	✓	✓						
Australian	✓	✓	✓	✓	✓	✓		
Egyptian	?	✓	?		✓	?	?	?
Finnic	?			✓	✓		✓	
Oceanic	?	✓	✓				✓	
African		✓						
Japanese			✓				✓	
—								
Atlantis Myth	✓		✓	✓		✓		✓
'Libyan'	?		✓	✓		✓		✓
Biblical	✓	✓	✓	✓	✓			

A similar fate befell the myths of pre-Columbian America. Classical sources take us back to the early centuries BC, with some as old as 800 BC. Judeo-Christian tradition dates securely from the early centuries BC, as also do the Chinese and Indian pseudo-historical traditions.

Only Egyptian and Mesopotamian evidence, such as the Gilgamesh epic, may take us back to even older written sources. The oldest written source of all is the Egyptian pyramid texts. From the traditions of the Flood and the antediluvian paradise, we may conclude that any real world it remembers, and the catastrophic destruction of it, must date from *no later* than the fourth millennium BC. This is the era at which we should seek evidence of a cataclysmic flood event in human prehistory; although in all probability there have been many earlier episodes.

There has long been a tendency in western thought, to treat the Near-Eastern and Classical myths as somehow more authentic than other versions, much as all culture and civilisation was once deemed to have diffused to other parts of Europe (and even to the rest of the world) from an eastern Mediterranean germ. Such a view of the past is surely no longer tenable.

Are we to assume that ancient nations came together in some great Neolithic 'summit conference' to standardise all their myths and traditions, with the intention of leaving behind fabulous tales of catastrophic events – solely to confuse later generations? The notion is absurd! How then do we account for all the similarities between them? The degree of correspondence between the myths shows us that their ultimate inspiration lies in a very ancient and terrible reality.

References to Chapter One

1   The cubit was originally the length of the human forearm, or about 43-56 centimetres.

2   Genesis, 7, 11

3   Josephus, Antiquities of the Jews, 2I, 3

4   Sollberger, 1971, p 10

5   Jacobsen, T. 1939

6   Berossus, Book II, see Burstein, S.M. p 19

7   ibid, p 20

8   Sollberger, E, 1971, p20-21

9   Apollodorus 1,7,1-2; Ovid Metamorphoses, 1,318-415

10 From the translation of Sir J.G. Frazer.

11 Hesiod, Works and Days, 109-120

12 Ovid, Metamorphoses, 1

13 Apollodorus, 1, 6, 1-3

14 The description of the hairy giants gives away that this is an imagery of a sky filled with comets, as the name comet simply means 'long-haired (star)'.

15 Apollodorus, 1, 6, 3

16 Agni Purana, II

17 Purana of Vishnu, 6,4.1-12

18 Purana of Vishnu, 24, 25
19 Burstein, S.M. p 14
20 Werner (1932), preface, xiii
21 Ronan, C.A. The Shorter Science and Civilisation in China, 1984, vol 1, p 84.
22 Shu Ching, see the translation by Clae Waltham
23 For a discussion of Ragnarok and parallels in other mythologies, see Crossley-Holland, K, p 234-236
24 Myvyrian third series, triad 13, Translation by Davies, E., 1804, p157
25 Myvyrian third series, triad 97,
26 For a discussion of the authenticity of these myths, see the translations of Bromwich (1968). The present author is perhaps more tolerant of the authenticity of the underlying myths than are the Welsh scholars, both for the reasons stated in the main text and because of their compliance with the myths of other nations.
27 Strabo, 4,4,4
28 The Exeter Triad, see Bromwich, R, 1978, introduction, xc
29 See Miller & Taub, 1993, p68-69
30 Poignant, R, 1985, p 118
31 Williams, R, Bibliotheca Sacra, 1907, 64:148-167
32 ibid
33 Levi-Strauss, 1986, p184
34 Plutarch, Isis and Osiris, 367
35 Job, 41,1; on the equivalence of Leviathan with Apsu and Tiamat, see Sandars, 1971, introduction 63-65; see also Enoch, 60, where the quaking of the heavens at the Flood is caused by the movements of Leviathan and Behemoth.
36 Clube and Napier, 1982
37 See N.K. Sanders, 1971, p 26
38 These may indeed be the same creatures as those mentioned by Berossus, coming from the same source, the Babylonian creation epic Enuma Elish.
39 From the translation of N.K. Sandars
40 Nebiru was also sometimes a title of Marduk and of the planet Jupiter.
41 Enoch is unrecorded in the Old Testament other than as the great-grandfather of Noah.
42 Enoch, XXI,6; from the translation by R.H. Charles, p 46
43 ibid, LXXXII
44 ibid, LXXXIX,6
45 Josephus, Antiquities of the Jews, 3,1
46 Geoffrey of Monmouth, translation by Lewis Thorpe, p 184-5
47 The references to Saturn and the Pleiades are an indication that some parts of the prophecies are based on authentic ancient astrology. On this see chapter 5 below.
48 Geoffrey of Monmouth, translation by Lewis Thorpe, p 179
49 Diodorus Siculus, V.18.2. 19-21
50 See, for example, Julius Caesar's description of Britain
51 Diodorus Siculus, III,53-54
52 Diodorus Siculus, III,56,1-2
53 Apollodorus, II, v, ii
54 Utterance 519, §12, from the translation by R.O. Faulkner
55 Literally: 'happy fields', or 'fields of happiness'.
56 Homer, Odyssey, 560-570
57 Hesiod, Works and Days, 165-175
58 I have explored this connection more deeply in Chapter 13 of *The Atlantis Researches*.
59 Kalevala, Runo XVI,151-158, from the translation by W.F. Kirby
60 ibid, 170-180
61 Vita Merlini, translation by S. Evans, 1958, introduction, vxii

2

Catastrophists and Uniformitarians

It is a common maxim that once an idea finds its way into the textbooks then it can be very difficult to overturn it. Yet one need only pick up a fifty-year-old book on almost any subject to see how far ideas have moved on. How many of the scientific papers published today will retain their relevance in fifty years time?

Nowhere is this more true than in the fields of geology and archaeology. Fifty years ago the theory of continental drift was yet to be accepted; and palaeontology relied upon the former existence of 'land-bridge' continents: Lemuria, Mu and Atlantis, to explain the observed distribution of fossils. Modern theories of plate tectonics and vulcanism have swept all this away. Fifty years ago too, archaeologists were confidently cross-dating various human cultures against the (still by no means proven) chronology of ancient Egypt. These certainties too were swept away by the advent of radiocarbon dating and then once again by tree-ring calibration of those dates. Now these new certainties are as firmly entrenched as were their predecessors.

So before it is possible to make a case for catastrophism in human prehistory one must first comprehend how the currently accepted norms of these two disciplines have come about.

Catastrophism, in western science at least, has its roots in the Biblical teaching of the creation and Noachian Deluge. Thus, the creation of the world and its regeneration consequent upon the Great Flood was largely a matter of medieval faith in the scriptures. The world developed this way because a supernatural creator had deemed it so and therefore further enquiry was neither necessary nor desirable. In 1654 the Archbishop of Armagh, James Ussher used a literal interpretation of the Bible to derive a date of 4004 BC for the creation; and this chronology was subsequently embodied in many printed editions of the Bible. The belief that the world was just a few thousand years old therefore became firmly entrenched in seventeenth century thinking.

The awakening of science, particularly of physics and astronomy, during that same century brought new ways of thinking. The invention of the telescope and the work of Galileo (1564-1642) had replaced the ancient geocentric view of the solar system and opened the way for Newton to explain planetary motions. Educated men of science began to ask how the traditions of a Deluge might similarly be explained in rational terms.

The Reverend Burnet in his *Sacred Theory* proposed that the Earth must have been created with its axis vertical, not tilted, and that before the Deluge there were no seasons; but then waters had 'boiled-up' from the abyss below the Earth and the land sank into it, leaving just a few continents and islands.[1] He left open the cause of the Deluge. William Whiston, who carried forward Newton's physics, was the first to propose a *comet* as the agent, which by passing close to the Earth, had drawn the waters of the Deluge from the Abyss below.[2]

The constraints of religion continued to hold back geological theorists. In Italy Nicolaus Steno proposed that the layers or *strata* of marine fossils in his native Tuscany could only have been created by two separate floods.[3] The first he presented as a record of the original creation of 4004 BC; and the second 1650 years later as the effect of the Deluge. His work is the foundation of modern sedimentary geology. In England, Robert Hooke, perhaps better known for his law of elasticity, was also working on the subject of fossils and floods. He argued that fossilised creatures buried deep within the ground could not all result from a single flood.[4]

For much of the next century, the presence of fossils in the sedimentary layers continued to cast doubt upon the religious interpretation of the age of the Earth. The painstaking study of rocks culminated in Werner's classification of rock types.[5] But it was the work of James Hutton that really tested the short time scale of the creation.[6] He saw geological processes as driven by heat within the Earth: sediments washed into the sea by rivers were raised-up as new land by the Earth's internal heat, as evidenced by volcanoes and earthquakes. Although he did not openly challenge the religious chronology of the day, it was clear to many people that such theories required the passage of an immense span of time. It was not long before Christian fundamentalists began to attack Hutton's theories.

There grew then, to be a distinction between the mere accumulation of erratic boulders and the comparatively fresh bones at various places; and the truly ancient sedimentary rocks containing fossils. The former might still be viewed as the result of the Deluge, but the latter demanded a much greater expanse of time. Thus in the early nineteenth century the Noachian Flood still retained a place in geology.

When Charles Lyell's *Principles of Geology* was first published in 1830-33 it was a total rejection of all things catastrophic. As the sub-title of his work openly stated, he sought to explain all geological phenomena 'by reference to causes now in operation'. His 'uniformitarian' hypothesis retained no place either for the Deluge or for Hutton's heat engine. The world we see around us came about by only the action of weather and erosion, together with occasional vulcanism and earthquakes, over periods of time so vast that the human mind cannot readily encompass them. Lyell did not challenge religious creation, or offer any view of the age of the Earth, but merely argued that since the creation the causes of change have always operated as we see them. Lyell's *Principles* continued to form the basis of geology for the remainder of the nineteenth century, going through several editions up to his death in 1875.[7]

What then of Diluvialism? Before Lyell it had been the norm to explain any out-of-place rocks or bones as the residue of the Biblical Flood. William Buckland (1784-1856) had argued that the widespread beds of gravel, or 'alluvium', found in many low-lying areas, together with the skeletal remains of diverse creatures found in caverns, had all been deposited by the waters of the Deluge.[8] These ideas were largely refuted by the Swiss scientist Louis Agassiz, who recognised much the same phenomena in the terminal moraines of Alpine glaciers. He proposed a former 'ice-age', when Alpine-style glaciers and ice-caps had covered much of the northern hemisphere.[9] He spent much of the next thirty years touring Europe and America in search of evidence for former glaciation. Lyell embraced the ice-age theory without difficulty; he had himself suggested that many of the erratic boulders may have been trapped within icebergs and drifted to their present positions – hence he termed such deposits 'the Drift'.

Charles Darwin was greatly influenced by Lyell, reading him avidly during his long voyage on the Beagle. Without Lyell's long chronology for geological time, he and his contemporary Alfred Wallace could never have formulated the theory of evolution. The religious controversy generated by his *Origin of Species* is surely well known by everyone, but over the course of the nineteenth century both it and Lyell's geology came to be accepted.[10]

Thus, there was no longer any place for catastrophism in scientific thinking. The world and its various life forms could all be satisfactorily derived by gradual processes; and the more recent geological anomalies explained as the effects of former ice-ages. By the beginning of the twentieth century, to uphold any form of catastrophism was looked upon as outmoded and unscientific. The Biblical account of creation and the Deluge receded to the status of religious symbolism, upheld by only the most ardent fundamentalists. This was the view that western science gave to the rest of the world during the era of colonial expansion. At the same time missionaries were forcing Christian teaching upon other cultures – many of whom held traditions that had never had any difficulty in visualising the very great age of the Earth.

Ice Ages

In a modified form, Agassiz's ice ages were perfectly consistent with the uniformitarian principle. Since catastrophism was now discredited as a cause of geological phenomena, so it could no longer to be seen as an active agent of change during *human* prehistory. Glaciology was the new champion. The rise of human civilisation has taken place only since the last retreat of the ice. Although here we are primarily concerned with catastrophic phenomena during human prehistory, some discussion of the ice age and its associated climate changes is a prerequisite to understanding all that has happened since.

By the 1890's, geologists in America and Europe were able to map the extent of the northern Drift deposits. It could be seen that great ice sheets had formerly covered most of Canada and the USA as far south as the Great Lakes, with another ice sheet centred on the Rocky Mountains. In Europe, a similar sheet had enveloped Scandinavia and northern Britain, with another extending from the Alps. All these could be seen to have retreated, leaving only the Greenland ice sheet; and of course, the Antarctic ice sheet, seemingly little altered since the age of ice.

It soon became apparent that, as a consequence of so much water being abstracted from the oceans to form ice caps, then the world-wide sea level must have been much lower during the ice age. At first the occurrence of raised beaches and cliffs at various places were something of an enigma, as they seemed to indicate that the former sea level had been higher rather than lower – for example in Scandinavia sea shells are found at altitudes now in excess of 1000 ft (300m). It became clear that the weight of ice must have been so heavy, that it had depressed the crust.[11] Therefore, the melting of the ice sheets at first raised the sea level, followed by a more gradual rebound of the formerly glaciated land to the heights we now see.

How many times have you read, in one context or another, the expression, "after the glaciers melted the sea levels rose…"? This is the start of many circular arguments that are entrenched in uniformitarian thinking.[12] The various estimates of the size of the former ice cap have been largely derived by calculating the weight of ice supposedly needed to depress the land by the extent observed. The sea level rise proves the existence of former ice sheets – yet it is the variation of the ice sheets that is advanced as the cause of the observed sea level changes.

From the 1840's onwards a variety of theories were proposed to explain the cause of ice ages. The French astronomer Leverrier had shown that the Earth's elliptical orbit was subject to perturbation by the other planets over a period of 100,000 years, from near-circular to a maximum eccentricity of about 6%.[13] This phenomenon, combined with the *precession of the equinoxes* – known since ancient times – over a period of 26,000 years, gave a possible cause for long-term climate change. The pull of the Sun, Moon and planets on the Earth's equatorial bulge also cause the axial tilt to vary between 22° and 25° over a cycle of 41,000 years.

The astronomical theory proposed by Croll relied principally upon the orbital eccentricity and failed to satisfy the critics.[14] Nineteenth century science was quite incapable of assigning an accurate date either to the ice age or to its end and another hundred years would elapse before an astronomical theory of ice-ages could be adequately verified. Croll's theory was revived in a more comprehensive form by the Yugoslav Milutin Milankovitch in the years up to World War II. However, it the accumulation of sufficient fieldwork and the advent of radiocarbon dating was required before his astronomical theory was finally proven to the satisfaction of the geological community.[15]

Twentieth century research charted the advance and retreat of the ice caps over the millennia. The combination of the various astronomical factors was able to confirm the sequence of alternating cold periods and warm 'interglacial' periods. Lyell had referred to the age of ice as the

Pleistocene epoch but subsequent geologists have modified his term to encompass the entire period from the end of the *Pliocene* (2 myr ago) to the end of the last glacial period. The roughly ten-thousand-year period since the last ice retreat is known as the *Holocene* epoch. Geologically, this short period, which encompasses the entire period of human civilisation, is revealed as really no more than the most recent of many interglacial stages.

Archaeologists are always right!

Modern archaeology sits on top of an assumption of uniformitarian geology. As a division of academic study, it has evolved from its antiquarian roots. Just as with geology, the Bible determined the historical time scale. Realistic enquiry into human prehistory could only commence with the realisation that the world was much older than a literal interpretation of the scriptures would allow.

The Bible stipulated that only Noah and his family survived the Flood; and so all genealogies, especially royal ones, were incomplete until some way could be found to trace descent all the way back to the Ark![16] Even as antiquarian enquiry emerged from these medieval constraints, we may still see the influence that religion imposed upon the early attempts to explain ancient monuments and artefacts. As recently as 1804, for example, the Reverend Edward Davies, would propose that the ancient Celts and their Druids were descended from one of the lost tribes of Israel.[17] Indeed, in 1802 William Paley, in a work entitled *Natural Theology*, would reassert the view that the stories in Genesis, including the Flood of Noah, were a historical fact.

Much of modern archaeology, especially Egyptology, has its roots in the search for evidence in the former Bible lands, to prove the New and Old Testament stories. It was Napoleon's interest in Egypt that brought the first detailed survey of that country's monuments; indeed much of the work of the French Academicians, including the famous Rosetta Stone, ended up in the British Museum when they were brought back to Britain as prisoners. Scholars were spurred to decipher the ancient hieroglyphic texts and, in 1822, Champollion recovered the secrets of the hieroglyphs that had been forgotten for two thousand years. Along with this knowledge came a realisation, that Egyptian civilisation was incredibly ancient. The French pioneer work gave a huge impetus to European interest in Egyptian civilisation; and many expeditions to Egypt and the nearby Bible lands were privately funded. The ancient tombs were opened up, as much in search of gold and treasure as any pretence of science.

That humankind had existed before the Flood was evidenced by the discovery of artefacts such as flint hand-axes and bone tools in the so-called 'antediluvian' deposits. The geological revolution brought in its train a realisation that such artefacts must be far older than biblical chronology would allow. The reappraisal began with the work of Lyell and Darwin, although it was not until 1871 that Darwin felt confident enough to propose that humans were descended from apes, rather than a creation of God in his own image. The mould was finally broken

A Danish museum curator, Christian Thomsen was the first to classify human artefacts into three ages; and his idea that human development began with a Stone Age, followed by a Bronze Age and an Iron Age was in widespread use by the mid-nineteenth century. A decade later the Stone Age would be further divided into the Palaeolithic (Old Stone Age) the Neolithic (New Stone Age), with the recognition that the Palaeolithic extended far back into the ice ages. The Biblical chronology was thus gradually pushed out of archaeology; and along with it went any belief in the Flood as a historical reality.

Nineteenth century science possessed no way to assign calendar dates to artefacts and monuments unless they could be associated with a historical culture. All that could be done was to classify the axe-heads, pottery, arrowheads and so forth, into some kind of evolutionary sequence – a method known as *typology*. Thus, a monument with primitive artefacts could be deemed older than one with a more developed style. Sometimes pottery or tools from a neighbouring culture could be found at a site, thus establishing them as contemporary. In this way, a whole series of 'floating' chronologies grew up, by which ancient cultures could be compared, but with no way to assign them a precise date. By the close of the nineteenth century, the meticulous standard of documentation adopted by Pitt-Rivers in his excavations of the barrows of Dorset, had set the standard for all subsequent archaeological sites. The antiquarians had become archaeologists.

The problem of chronology remained; and it was Egypt that supplied a way forward. Egypt, and to a lesser extent neighbouring Mesopotamia, were historical cultures that overlapped with Greece and Rome, for which reliable dates can be obtained from classical history. The deciphering of Egyptian hieroglyphs and Mesopotamian cuneiform inscriptions revealed a number of king lists, which potentially extended back for thousands of years. If excavated finds could be positively associated with the reign of an Egyptian king, then its era could be established by counting-back the reigns from a historically fixed date. The more and older these 'fixed dates', the more confident Egyptologists could be in the derived chronology. Furthermore, if similar Egyptian artefacts could be discovered in adjacent countries then their own floating chronology could be tied into the history of Egypt. This *cross dating* technique would carry forward archaeological thinking for the next half-century.

Alas, and not for the last time, the Egyptologists could not agree on precisely what the historical chronology of Egypt should be. The basis of the chronology was a king list of some thirty 'dynasties' preserved in the chronicle of the Egyptian priest Manetho (c.250 BC) together with a number of other king lists recovered from Egyptian excavations.

In the early twentieth century, various estimates for the date of the Egyptian First Dynasty were championed by different Egyptologists. Those who treated the dynasties as consecutive, favoured a date as early as 5867 BC, but others who treated some dynasties as reigning in parallel have produced dates as low as 2950 BC.[18] The various 'fixed dates' established by the Egyptologists tended to favour the shorter chronology. The long-accepted standard in *The Cambridge Ancient History* assumes a date of 3110 BC for the commencement of the Dynastic era – but this is far from being the final word on the subject! This date was crucial if the Egyptian chronology was to be the yardstick to date other cultures, in Mesopotamia, Europe and even further afield.

Thus, the early twentieth century saw a debate between the *evolutionists* and the *diffusionists*. The former saw civilisations as developing independently in various parts of the world; the later believed that discoveries such as agriculture, the use of bronze and iron, even esoteric advances such as mathematics and astronomy, could have been invented but once. From this cultural germ, ideas then 'diffused' to other parts of the world. The champion of diffusionism was Vere Gordon Childe, the first professor of archaeology at Edinburgh University in the 1930's. His influential theories, fuelled by a confidence in cross dating methods, saw all culture emanating from Egypt and the Fertile Crescent. From there, ideas reached out to Stone Age Europe and beyond, in an event that he termed the *Neolithic Revolution*.

Ironically, it was archaeology that would inject renewed vigour into the story of the Flood, in the form of a clay tablet discovered in the ruins of Nineveh. George Smith, an Assyriologist at the British Museum in 1872, had been given the task of translating the tablet. Half-expecting that he might find a version of the Flood story, he was no less elated to find there the story now known as the *Epic of Gilgamesh*. Its details so closely parallel the Biblical version that there seems little doubt that both had a common origin in ancient Sumeria.

Scholars had long known of a Greek text called the *Babyloniaca* or *Chaldaica*, written by *Berossus*, a priest of Bel-Marduk at Babylon in about 280 BC. Little is known of him, but his book was also known to Manetho, who similarly believed that the Flood, whatever form it may have taken, occurred just before the first dynasty of Egypt. Other Babylonian king lists similarly tell of dynasties of kings "before the Flood" and it is evident that the people of ancient Mesopotamia believed in the universal Deluge as a historical reality. Here then is the first piece of historical evidence that dates a flood event to the late fourth millennium BC – a subject to which we shall return.

Between 1929 and 1934 a joint American and British expedition to Iraq under Sir Leonard Wooley carried out deep excavations at the site or Ur – the city from which Abraham came. This remarkable expedition is surely one of the landmarks in archaeological history. In deep test pits near the wall of the ancient city, a thick layer of silt was discovered beneath several layers of human occupation. Wooley immediately declared it evidence of the Biblical Flood and ordered the workers to continue digging. Beneath the silt lay older layers of occupation. The city apparently had been rebuilt after it had been buried by a catastrophic river flood. Similar deposits have been identified at many other ancient Sumerian cities such as Kish, where it can be dated to the Mesopotamian Early Dynastic period.

Thus, for twentieth century archaeologists, the origin of the Flood story could now be assigned an approximate historical date, yet the Flood itself remained (and remains) just a myth!

The Mesopotamian chronology was something of a triumph for cross dating. At Jemdat Nasr near Baghdad distinctive cylinder-seals were found, together with a characteristic style of pottery. The seals were intended to be rolled-out onto clay tablets to impress a pictographic symbol – an early form of writing. These seals have also been found in Egypt at sites that predate the Dynastic era, where they appear to have been traded as ornamental amulets. Many other similarities have been noted between Sumeria and predynastic Egypt. Thus, using local typology we may assign the so-called 'Flood' deposits to a date around 3100 BC.

By a myriad of similar correspondences the floating chronologies across most of Europe were tied into the near-eastern chronology, all of it dependent upon the accuracy of the Egyptian king list and the few specialist Egyptologists who determined it. Textbooks were printed and students were taught by their professors. Archaeological certainty was set almost literally in tablets of stone.

Archaeology was put on a more scientific base after the Second World War when the Radiocarbon dating method was developed by Professor Libby of Chicago. Archaeologists could now take samples of wood or bone from their excavations and have them dated in a laboratory. Radioactive elements decay at a predictable rate and the amount of carbon-14 present in the atmosphere could be readily measured. It seemed that to derive the date of an artefact was now simply a matter of measuring the amount of carbon-14 in the sample and then reading off the date from the decay curve. This method provided the first independent check upon the cross-dated chronology, but at first, archaeologists were reluctant to publish the new dates unless they supported the accepted wisdom.

As more and more radiocarbon dates accumulated during the 1950's archaeologists were able to adjust many of their theories or, for a while, to fudge the difficulties. At first, samples of known age could be dated and the method proved valid for the sites of classical Greece and Rome. However, when radiocarbon dates were tried for older sites in Egypt, and for the cross-dated cultures throughout Europe that depended upon it, then they often appeared younger than expected. Before about 1000 BC the dates became increasingly divergent from the established chronology, and many Egyptologists began openly to challenge the credibility of the technique. Radiocarbon dates for the Egyptian Old Kingdom, which the king lists put at around 2500 BC, were coming out with a radiocarbon age of less than 2000 BC. The Egyptologists refused to revise their historical chronology.

Another dating technique, dendrochronology, provided the answer. In theory, the age of a tree can be determined by counting back the annual growth rings from a known date. As the climate varies, some years show a thick ring of growth, while years with a poor summer will show a thin ring. The sequence of these thin and thick rings can be recognised by the specialists. In many parts of the world, overlapping sequences of rings have now been painstakingly compiled from ancient trees preserved in waterlogged ground. Wood from each growth ring can then be radiocarbon dated and compared with the tree ring date, to give a calibration curve.

In the 1960's however, it was wood from the long-lived Californian bristle-cone pine tree that provided an unbroken sequence of tree rings back beyond 3000 BC. Radiocarbon dates from the bristle-cone pines proved conclusively that the discrepancy in radiocarbon dates was real. Before about 1000 BC the atmosphere must have contained more radiocarbon than it does today; and this was making the older artefacts appear young. Moreover, the calibration curve is far from smooth and for some periods, this can give rise to two (or more) possible ranges of dates for a sample. Together with the inherent statistical uncertainty, this means that radiocarbon dating is far from an exact science.

Nevertheless, the calibration technique was good enough to prove that the trusted Egyptian king list was closer to reality than uncalibrated radiocarbon dates. The real shock came when artefacts that had been dated by the cross-dating technique proved to be earlier than the archaeologists believed. In particular the Neolithic and Bronze Age cultures of Western Europe were shown to be older than the civilisations in Egypt and the Aegean that supposedly had influenced them.

Before the radiocarbon revolution, anyone who had proposed that the megaliths of Europe were older than the pyramids of Egypt would surely have been pounced upon by the archaeological establishment. The theory of diffusionism was crumbling and the edifice of cross-dating was shaken to its to its foundations. Who within the archaeological community would dare to give it the final push? In 1973, Professor Colin Renfrew stuck his head above the parapet and restored order. As he said in the introduction to his *Before Civilisation*: "Archaeologists all over the world have realised that much of prehistory, as written in the existing textbooks, is inadequate: some of it quite simply wrong."

Has the new chronology therefore undone all the errors and prejudices of earlier thinking? Far from it. The attitude of archaeologists to folklore, myths and legends remains as dismissive as it ever was. If the indigenous folklore of a region contradicts the archaeological consensus, then archaeologists ignore it, just as they once ignored radiocarbon dates.

Thus, it may be seen that modern chronology rests upon a sequence of assumptions about the past that have come about by a steady refinement from a nineteenth century base. Along with the defeat of Biblical chronology went the belief in the reality of the Deluge and many other miraculous events in religious sources. As the saying goes, the 'baby', was thrown out with the bath water! Truth about the past is now to be determined solely by whatever can be dug from the ground and dated by radiocarbon. Only 'hard' evidence counts. Whenever myths and legends differ from the received wisdom of the archaeologists then the archaeologists are always right. One prominent archaeologist once suggested that *only* professional archaeologists should be allowed to propose theories about the past.[19] Can we really leave the final word about prehistory to a profession that has thrown away its own certainties twice in half a century?

A Universal Deluge – or a little local difficulty?

Of course, belief in the reality of a catastrophic flood in prehistory did not cease just because the geologists and archaeologists went their own way. Religious fundamentalism and devout belief in the literal word of religious scriptures has never entirely gone away.

Even before the triumph of uniformitarianism many explanations for the Flood were proposed, by authorities who could truly be called 'scientific'. It is instructive to examine why these theories failed to take hold. Some of the most interesting views came from mathematicians who carried forward the work of Newton and applied it to planetary orbits and rotational dynamics.

A fundamental tenet of Newton's physics was that the mass of a body could be represented as if it were concentrated entirely at its centre of mass. This led the way for Laplace and Poisson to describe why the Earth's shape is a flattened ellipsoid as a consequence of its rotation. A sphere may rotate about any arbitrary axis and its axis is completely unstable, but a flattened spheroid such as the Earth has only one stable condition, and that is when it rotates about its minor axis. Although the Earth's rotation is stable about the north–south axis, the *axis of figure*, it could temporarily wander away from this stable position if the shape of the planet were to deform. The ellipsoid shape of the oceans must then immediately conform to the new dynamics, while the figure of the 'solid' earth remains allied to the figure axis. The ebb and flow of the oceans at the continental margins would therefore resemble a giant tide; indeed modern geophysics would call this phenomenon a *pole tide*.

The French mathematician Pierre Simon Marquis de Laplace (1749-1827) in his work on celestial mechanics considered what might happen if the rotational poles wandered on the surface of the Earth. He saw in this a mechanism that could explain the Biblical Flood.[20] Laplace thought it possible that the strike or even a close approach by a comet could cause the seas to spill over the land, drowning mankind in the universal Deluge and bringing about the extinction of entire species:

Laplace himself computed, based upon the orbit of Lexell's comet of 1780, that cometary masses could be no more than 1/5000 part of the Earth. Later astronomers would compute the orbits of other comets and estimate their true mass. This would show that most were but a few kilometres in diameter and therefore far too insubstantial to exert any tidal pull upon the world's oceans. Even a close fly-by or a passage through the luminous tail would have no effect upon the great mass of the Earth and certainly could not disturb the axis. Another astronomer would assert that even a direct impact by a comet would be no worse than being whacked by a huge cushion! By the mid-nineteenth century the belief that comets were ethereal bodies, which in any case seldom hit us, had become firmly entrenched. This dismissive attitude would persist until virtually the end of the twentieth century.

The mathematician Leonhard Euler studied the characteristics of a rotating ellipsoid and applied his findings to the motion of the Earth. His simple equation describes how the axis of rotation, once displaced from its position of natural stability, must describe a circle about the axis of figure in a period determined by the flattening.[21] It says nothing about the amplitude of the motion, which is set by the starting conditions; nor can it say anything about its cause. Given the amount by which the Earth is flattened at its equator then this period turns out to be about 305 days for a solid body of the Earth's dimensions and velocity of rotation. This is termed the *free nutation*, because it not sustained by any outside force. Perhaps then, a comet would not be needed to explain the Flood after all?

However, Euler's nutation applies only for a rigid body and the true natural period of the Earth's wobble is modified by other factors. Modern observations, as we shall examine in a later chapter, give the true period as 420-440 days. Although many nineteenth century researchers already knew that the interior of the Earth was far from solid, this did not prevent a battery of notable astronomers, from searching the astronomical data for some small evidence of the predicted ten-month Eulerian wobble. The eminent Lord Kelvin even announced to the British Association in 1876 that he had found the free nutation. He had not.

Nonetheless, the principle was understood by many astronomers and geologists and the concept of a 'pole-shift' of some kind was seriously proposed. However, for nineteenth century men of science it was no longer the Flood that required a catastrophist explanation, but rather the anomalous fossil evidence and the climate changes associated with the newly postulated ice ages; and these were already perceived as too early to have played a role in human history.

Fossils of temperate and tropical creatures were increasingly being discovered at high latitudes where today the climate is too cold to support them. This suggested that the rotational poles could have wandered over the course of geological time. The discovery too, of misplaced bones and frozen carcasses of mammoths and other 'antediluvian' creatures in the Drift deposits led to suggestions that a more rapid shift of the poles, or a change in the tilt of the axis, must have brought about their demise.

An early opponent of this theory was British Astronomer Royal, Sir George Airey, perhaps better known for his role in the discovery of the planet Neptune. Airey was noted for the meticulous recording of his observations, sometimes to the point of absurdity. A colleague joked that if he even used a sheet of blotting paper in the course of his work, then he would annotate its use and file it away in his notes!

In reply to a published letter by his friend and geologist Sir Henry James, Airey argued that the popular notion of a pole shift could not possibly deliver the scale of effect that the geologists required. Geologists, he protested, did not appreciate the huge forces involved. His words neatly illustrate the attitude of subsequent mathematical astronomers and geophysicists whenever the idea of a pole shift has been proposed:

> I do not question the principle of Sir Henry James' speculations; but I very greatly doubt the adequacy, in magnitude, of the cause to explain the proposed effects.

And this, as we shall see, is indeed a powerful objection. Airey calculated that even the raising of a vast mountain range, by some explosive volcanic event, could not displace the rotational poles by more than a few miles on the surface.

A related theory was that of Henry Howarth who, in a series of papers in the journal *Nature* during 1871-2, proposed that all the circumpolar lands had only recently emerged from the sea.[22] Taken at face value such evidence suggested that the Earth's rate of rotation had very recently increased, causing the bulk of the oceans to migrate towards the equator. The arguments against a recent change in the length of day are much the same as those against a pole shift. Much of Howarth's evidence would today be dismissed as due to local movements, or the rebound of formerly glaciated regions since the ice age.

The new geology also brought with it a problem for evolutionary biologists. How had the various species migrated, as they evolved, from one continent to another? The zoologist Philip L. Sclater proposed a sunken continent called *Lemuria* to explain the presence of fossilised lemurs in both India and Africa. He suggested that a 'land-bridge' continent had once existed in the Indian Ocean, linking Africa, Madagascar and India; and that this had somehow subsided away. Similar continents would soon be postulated in the Pacific and Atlantic oceans; and had not Plato recorded the submergence of just such a mid-Atlantic continent only ten thousand years ago? Nineteenth century oceanography was not capable of proving or disproving such a theory.

In the early twentieth century a new theory, that of continental drift, began to erode the credibility of both the polar wandering and sunken continent hypotheses. If the poles could not move, then perhaps the continents had migrated around the world. When Alfred Wegener's theory of continental drift was first published in 1912, it was derided in the customary manner. One of its most ardent critics was Sir Harold Jeffreys, a pioneer of mathematical geophysics and seismology in the twentieth century. He never accepted the theory of continental drift, adamant that the mantle was simply too rigid.[23]

The steady advances in oceanography throughout the twentieth century saw the continental drift theory overtake its rivals and develop into the modern science of plate tectonics. Here was an acceptable uniformitarian process that could explain all the geological anomalies by *gradual* movements over vast periods of time. It can be seen that over millions of years, the continental plates have migrated due to volcanic processes at the mid-ocean ridges and troughs. Other strong arguments against polar wandering would come from the study of the Earth's magnetic field, which suggested that, the geographic and magnetic poles must have remained sensibly close to their current geographical positions throughout geological time.[24] However, this cannot preclude a wandering of the rotational poles by a few degrees of latitude around their present positions.

Oceanography had therefore disproved the sunken continent hypothesis and polar wandering was no longer needed to explain the anomalous distribution of fossils. Sea level and climate changes could be explained either as the effects of the ice age or of its sudden end. As long as no one looks too deeply into the circular arguments, then a uniformitarian solution suffices for everything. The physicists' objections to changes in the Earth's rotation as the agent of geological change therefore ceased to be of any further relevance. The polar wandering theory was never disproved – it merely lapsed into disuse.

The Heretics

The idea of the Deluge, or of some other great destruction in human prehistory, is certainly a compelling one. So despite the academic consensus it is hardly surprising that so many people still wonder if an ancient reality lies behind the various myths and legends. The public are not scientists. Many enthusiasts fail to grasp the scientific arguments and others just do not want to try. Quite simply there is a *market* for simplistic catastrophism.

As was discussed in the first chapter, classical authorities and myths from around the world offer many stories of lost lands and ancient paradises. So long as the religious chronology prevailed then these could all be tolerated as memories of an antediluvian reality. Had not Plato, for example, hinted that his lost continent was submerged in the flood of Deucalion? And was this not the same flood as that of the Bible? As the Flood itself faded into myth, so the stories of St Brendan's isle, The Islands of the Blest, Dilmun and Atlantis were also left behind as anachronistic tales. How might these too be explained by uniformitarian processes?

The first successful attempt to explain an ancient cataclysm by science – and what a commercial success it was – can be seen in Ignatius Donnelly's book *Atlantis: The Antediluvian World*, published in 1882. Donnelly was not a scientist, but a retired US congressman. However, there can be no doubt that he possessed a vast general knowledge and was well versed in the science of his day. Perhaps the most lingering aspect of his work was his belief that there were many similarities between ancient Egyptian culture and the pre-Columbian civilisations of America, citing pyramid building, embalming, and the 365-day calendar, among others. He argued that the two cultures descended from a common origin: the survivors from the sunken mid-Atlantic continent.

Donnelly was the first of the 'mid-Atlantic' authors. The secret of his success was to give the American continent a share in the history of the Old World and to bring it home to the vast American audience. Many of his theories can easily be challenged today, but he had identified a formula that has been imitated by authors and publishers right up to the present. In Donnelly's day, no one knew how old the pyramids were. Another hundred years would elapse before oceanography would disprove the existence of land-bridge continents; and before radiocarbon dating would conclusively show that ancient Egyptian pyramids were three thousand years older than the pyramids of Mexico. Donnelly's hypothesis was perfectly sound by the science of the late nineteenth century. The more modern proponents of the 'lost civilisation' theories have simply never moved on.

The new factor in late nineteenth century catastrophism was that it was based on science rather than religion. Donnelly employed a volcanic explosion to sink his mid-Atlantic land-bridge. No longer was the Flood a universal catastrophe, but a little local difficulty, neatly lost in the Atlantic Ocean. Worse still, Atlantis became the ultimate cradle of civilisation. Just as the archaeologists would hold, that civilisation had begun in Egypt, so the Atlantologists would claim that it arose in Atlantis and the survivors of the disaster carried it to Egypt and America. Diffusionism was orthodox archaeology taught in universities for half a century, while Atlantology is a pseudo science disparaged by scholars. Yet, both theories share the dubious equality of being completely wrong.

This kind of pseudo-science was picked up by other authors, such as James Churchward, a former Bengal Lancer, who proposed a similar lost continent called *Mu*, in the Pacific Ocean. The extent of Mu encompassed most of Polynesia and supposedly sank 12,000 years ago amid earthquakes and tidal waves. Although written in the 1880s the first of Churchward's books were not published until 1926. His stated source was a series of Hindu tablets that he had seen in Tibet, but unlike Lemuria and Atlantis, no one has ever suggested a scientific basis to it. Polynesian mythology knows nothing of such a continent and stresses the emergence of the islands, rather than the reverse. Such minor considerations did not prevent publishers from accumulating profits from the late colonel's books.

Even the innocent Lemuria was appropriated and pressed into service as the ancient home of mankind. The German scientist Ernst Haeckel proposed it as the earliest home of the human species, whence the idea was picked up by various occultists. Madame Helena Blavatsky, founder of the Theosophical society, built Lemuria and Atlantis into her pseudo-history of mankind. These occult ideas flourished in Germany and eventually found their way into Nazi ideology where Atlantis became the original home of the 'Aryan' race. All of these ideologies required a *catastrophe* to explain the destruction of the ancient homeland – and there is surely no more certain way to discredit any hypothesis than to have its name linked with Hitler!

The astronomical nature of the ancient catastrophe was also explored by a number of pseudo-scientific authors. One such was the Austrian cosmogonist Hans Hoerbiger in his *Glazialkosmogonie* of 1913. He was even prepared to invent his own physics to compel the planets to move as he wanted them to move. He considered interplanetary space to be filled with a tenuous medium that deflected the planetary courses over long periods; and that one such planet was our own moon, only recently captured by the Earth.

Another German, Edmund Kiss, seized upon this cosmology to explain the 'calendrical' carvings on the great Gateway of the Sun at Tiahuanaco, Bolivia. This monument has had a special place in the history of catastrophism. Kiss proposed that the gateway and city of Tiahuanaco were incredibly ancient, and that the 'calendar' was based, not on the appearances of our own Moon, but upon the movements of a mysterious cosmic body that no longer exists.

These bizarre ideas were further popularised by the English author H.S. Bellamy in his books *Built before the Flood* and *Moons, Myths and Man*. On the 'evidence' of the calendar, Bellamy contended that this mysterious pre-Moon had orbited the Earth in a geostationary orbit above South America and thus attracted the oceans toward itself in a huge tidal bulge. In this way, he could place Tiahuanaco and the Altiplano at sea level and yet avoid the obvious counter that this tidal bulge must also have swamped Africa and Asia. The Flood of Noah and other cataclysmic events such as Atlantis were to be explained by the capture of our Moon, which had ejected the old moon and released the suspended waters to flood the Earth.

These works make the most painful reading today, for anyone with even a basic knowledge of astronomy or physics.[25] Modern radiocarbon dates have shown that the ruins on the Bolivian plateau are not ancient and date from the second to thirteenth centuries AD. Neither can it be proven that the symbols on the gateway represent a calendar at all.

Another catastrophist with an astronomy all of his own was Immanuel Velikovsky. His *Worlds in Collision*, published in 1950, proposed that the planet Venus was originally a rogue comet emitted from Jupiter, which had settled in its present orbit only some 4,000 years ago. For reasons that are difficult to explain Velikovsky's book produced a furore of academic scorn; his ideas were so outlandish that astronomers condemned them with a near-religious fervour, almost as great as that faced by Darwin and Hutton a century before. The negative academic reaction persuaded his publishers to drop the title, after which it became a best seller with another publisher. The controversy only served to encourage more people to read it and gave a notoriety to his subsequent books that they probably did not deserve.

Velikovsky, a Russian by birth and with a background in medicine and Biblical history, seems to have been inspired by some of the miraculous stories in the Bible. One example from the Book of Joshua tells us that the Sun 'stood still' for a period of a whole day.[26] This he took to indicate that the rotation of the Earth must have paused; and that such an effect could only have been caused by the close approach of another astronomical body of considerable gravity. In support of this hypothesis he drew upon a variety of legends from around the world and linked them all into his Venus hypothesis.

In his later book *Ages in Chaos*, Velikovsky argued against the established chronology of Egypt and sought to bring forward many of the events of the Egyptian Middle Kingdom from 1450 BC to around 850 BC, with knock-on consequences for subsequent dynasties. Here he was arguing on ground much closer to home and in the era before calibrated radiocarbon his theories were less easy to dispose of. You may well ask what is wrong with that? "Very little", would be my reply, but would his theories have received the attention that they did, were it not for the absurd reaction to his earlier work? The issue seems to have been that he openly challenged the uniformitarian consensus of archaeology at the very era when it was at its most entrenched; and he supported his theories by a naive astronomy that was such an easy target.

It is Velikovsky therefore, whom we must hold primarily responsible for the low standing that catastrophist research has subsequently gained among orthodox academics. Publishers and authors noted how well these books sold and consequently the number of books on ancient mysteries such as the pyramids, Atlantis, ancient spacemen and occult subjects has mushroomed since the 1960s. The concern for accurate science in most of these works is nil. This was accompanied by a retrenchment of the academic community such that it became near impossible to get even the most restrained of catastrophist papers past a referee.

No summary of modern catastrophism would therefore be complete without some mention of the role played by the publishing industry. The reader may think that it is an author who writes a book and then submits it to a publisher; and, if the work has merit then it will be published. Nothing could be further from the truth. Usually, publishers determine those subjects that can be profitably marketed. They then commission an established author – who may have but limited knowledge of the subject – to research and write a suitable book to specification. The book is then professionally illustrated and packaged to sell through distribution channels, to libraries and booksellers that specialise in that subject. Often success may be determined by the hyped image and reputation of the author, or upon whether the title is linked to a television series. True researchers are then obliged to read each of these glossy volumes, in case it should contain a paragraph of originality.

At the other end of the scale are the academic publishers. Here original works of research may be put out, often in low volume or at a loss, by universities; or by publishers who only consider recognised academic subjects. Here again, whether the book is published depends more upon the eminence of its author than upon the content. Alternatively, it may depend upon the patronage of a suitably distinguished referee. For a research paper to be published in an academic journal it must first pass the journal's referee, who is usually a senior academic steeped in current text book wisdom. Often it is only possible to smuggle a new idea past such a referee if you can cite a reference showing that someone else has already proposed your new theory!

Although the above review may be intentionally flippant, it does illustrate the fundamental problems faced by the catastrophist researcher – or indeed by anyone wishing to publish radical ideas. This polarisation of the debate ensures that any work on a catastrophist subject must conform to one or other of these roles.

During the later twentieth century, the contribution of Charles Hapgood is deserving of mention. In his books, *The Path of the Pole* and *Earth's Shifting Crust* (1958) he revived the concept of a pole shift as an alternative cause of the ice ages.[27] Citing various geologists of his era, he proposed that the Earth's crust as a whole could somehow slide over the mantle. He therefore explained the apparent expansion and contraction of the northern ice sheets by the sliding of the crustal shell across the North Pole, rather than by any movement of the poles themselves. He proposed that some 80,000 years ago Alaska covered the North Pole. From there the crust slid so that Norway assumed the polar position; Hudson Bay shifted over the pole at about 50,000 BP, with the crust slipping to its present position around 17,000–12,000 BP.

In his later book *Maps of the Ancient Sea Kings* (1966), Hapgood hit upon an idea, that has had a very long shelf life. He noted a surprising accuracy in the maps made by the Turkish admiral Piri Re'is in 1513, which predated most of the renaissance voyages of exploration. He suggested that this precocious geographical knowledge came from charts made by ancient navigators that had been preserved in the lost library of Alexandria. In particular, he proposed that the contours of the great southern continent were surprisingly similar to the outlines of Antarctica, as they would be if its ice sheet were removed. He therefore proposed that the maps must have been made by ancient mariners who had explored Antarctica before 10 000 BC when the supposedly ice-free continent lay further from the South Pole.

Perhaps because Hapgood's theory was so visual and supported by numerous maps and diagrams, it has been taken up by many other theorists. There is an obvious objection to it, however. It may be reasonable to suggest that a spherical shell could slide over a deeper sphere, but the Earth is not a sphere, it is a flattened ellipsoid that bulges at the equator by some 20 km (13 miles). Crustal slippage would therefore require a lubricating fluid layer, of similar thickness, at the base of the crust. Seismology shows no evidence for such a layer and the entire lithosphere is only 30–40 km deep. There is indeed a fluid layer, but this lies much deeper down, in the core.

Hapgood's work predated the final confirmation of the plate tectonic theory by oceanography and paleomagnetic evidence in the 1960s, which showed that true continental drift operates over millions, not mere thousands, of years. So again, crustal slippage was a reasonable theory at the era when Hapgood first proposed it. Despite the fact that the science has moved on, Hapgood's ideas have continued to influence many of the modern 'ancient mystery' authors.

It has been common practice (due to marketing considerations) for publishers to smuggle out books about Atlantis, ancient voyages and related 'new age' subjects, in the guise of books about ancient Egypt.[28] Another popular theme – as plate tectonic theories have made the mid-Atlantic land bridge theory completely untenable – has been to relocate the ancient cataclysm westward to the Caribbean islands. This has proven to be a very profitable theme, as it brought ancient Atlantis *even closer* to the vast American market for books. By citing Plato, this allowed ancient Atlantian mariners to have made transatlantic voyages, like those suggested by Hapgood, and it has even allowed the American continent to claim a share in the origin of Egyptian and Greek civilisation. This is a wonderful concept if you want to sell books in the USA and one has to buy and read these books to find out that there is absolutely no foundation for the theories.

The suggestion that the American continent was the lost continent referred to by Plato is as old as the European discovery of America; Francis Bacon proposed it as early as 1600. However, the first person credibly to propose that the islands of the Caribbean might be the surviving peaks of Atlantis, was the Scottish author Lewis Spence in the 1920's. For Spence, the cataclysm that submerged Atlantis (and which was remembered as the Biblical Deluge) was no more than the rapid rise of sea level at the close of the Ice Age. This, he believed, had flooded the low-lying area around Cuba and the Bahamas. Spence possessed an encyclopaedic knowledge of mythology and at the era when he wrote, the concept of a sunken mid-Atlantic land bridge was orthodox geology. The more recent 'mid-Atlantic' authors who have picked up this theme have no such excuse. [29]

The mass colonisation and diffusion of civilisation, from an American or mid-Atlantic 'lost civilisation' to the Old World, defies common sense. The occasional transatlantic voyage of exploration or trade, in either direction, should not be ruled out. However, can we really accept that huge conquering armies, led by kings, crossed the Atlantic over several generations, built an empire in ancient Europe and colonised North Africa as far as Egypt? If so, why do not all Europeans or Egyptians resemble native Americans? Where are their descendants and where were their cities? Why do not European languages resemble those of Central American tribes? Why were the tribes of the Caribbean islands and elsewhere, almost wiped out by Old World diseases, if there had been so much ancient transatlantic contact? The list of objections goes on. The theory fails the most basic scrutiny.

A typical recent offering was *When the Sky Fell* (1995) by Rose and Rand Flem-ath. Their view of a world struck by geological cataclysm would combine rapid continental drift with Hapgood's view of a sliding crust and an ice-free Antarctica – and carry it to its most extreme. After the cataclysm at the end of the Ice Age, or so they propose, the mid-Atlantic continent was set loose to drift around the globe. As the continents reconfigured, it came to rest over the South Pole where the lost city lies buried beneath the Antarctic ice! One would scarcely take this at all seriously, were it not for the fact that other authors have also proposed that Atlantis was Antarctica[30]. In the year 2000, the author Andrew Collins would record that this had become the most popular theory with enthusiasts on the Internet.[31] Come back Velikovsky. All is forgiven!

To find the reasons for this most strange development, we must examine some of the other books that were published towards the close of the twentieth century. In 1994 the author and television journalist Graham Hancock produced a best-selling book entitled *Fingerprints of the Gods*. His theme was not so much catastrophism, nor even Atlantis, but rather the former existence of a worldwide 'lost civilisation' that flourished during the Ice Age. Of course, a 'lost' civilisation requires a catastrophe of some kind in order to lose it.

Hancock built on this theme in his later book *Heaven's Mirror: Quest for the Lost Civilisation*, also in other books co-authored with alternative Egyptologist Robert Bauval, and in a much-repeated television series. He drew upon astronomical alignments and the layout of monuments such as the Pyramids of Giza, and Angkor Wat, as preserving ancient knowledge of the constellations as they would have looked 10,500 years ago. He also drew heavily upon the work of Bauval, who proposed that the three pyramids of Giza reproduce on Earth the alignment of the stars in the belt of Orion. [32]

Hancock proposed that around 10 500 BC the Earth endured "a series of cataclysms" and was in collision with several fragments from a disintegrated comet, recalled in mythology as a sky-monster. This all seems reasonable, as far as it goes, but what is the significance of the date

10 500 BC? Well, it is *approximately* the date that marks the end of the Ice Age and is close enough to the date given by Plato for the submergence of Atlantis. It was also the date that Hancock chose for his astronomical theories; given the rate that the Earth's axis precesses, this date gives the best fit between the appearance of the constellations and his chosen monuments. Where was the lost civilisation? Why, none other than the lost Antarctic-Atlantis!

Another 'alternative Egypt' author, Colin Wilson, has rather given the game away. He it was, who wrote an introduction to the Flem-ath's book and repeated the Antarctica theory in his own book *From Atlantis to the Sphinx* (1996). In his introduction, he reveals that the author, John West, suggested he speak to Hancock:

> We discussed my projected book on the Sphinx...He also threw off another name – Rand Flem-ath – who was writing a book arguing that Atlantis was situated at the South Pole. This made sense – Hapgood had argued that his ancient maritime civilisation was probably situated in Antarctica, and, now I thought about it, the idea seemed almost self-evident.

Wilson's revelations about Graham Hancock's work are even more revealing. Apparently a few months into his work Hancock had still not decided where his lost civilisation was to be situated. "The Antarctica theory came as a kind of deliverance", says Wilson. Apparently Hancock had received a letter of resignation from his researcher who had despaired of trying to find the lost civilisation demanded by his employer, "...as there was no known land mass in the world that could have accommodated such a civilisation". One wonders if Wilson ever thought how ludicrous these revelations would appear. Here we see catastrophism and alternative prehistory at its very worst! Books are designed to fill a marketing niche and the 'research' is tailored to fit. A circus of authors recommending and citing each other, each of them turning Hapgood's well-intended research into a kind of perverted science. The only person who comes out with any honour intact is the anonymous researcher who refused to have anything to do with it.

The New Catastrophism

In the present environment, it is therefore hardly surprising that orthodox academics have backed away from the idea of a catastrophe in recent human prehistory. The 'Atlantis' factor was a killer! Academics have to earn a living and their reputation can easily be destroyed if they stray off the tramlines. However, over the last twenty years and especially since 1995, there has been something of a renaissance for catastrophism. This has come about not so much by an opening of minds, as a reaction to some incredible events.

For geologists and palaeontologists, the recognition of the mass extinction of species during prehistory, has always been somewhat of an anomaly. Here was a fact, which, by its very nature, was not uniform. It may have been acceptable to suggest that creatures such as dinosaurs and ammonites could have died out gradually over say, a million years of climate change, but how could the same be said of mammoths, sabre-toothed tigers and dire wolves? These species all disappeared in the space of a few hundred years – and very recently in geological terms.

The theme that a comet or asteroid impact might have caused worldwide devastation at some point in the past is, as we have seen above, far from new. It was a resort of the earliest religious catastrophists and we may find it even earlier in the works of neoplatonists such as Proclus. Throughout the twentieth century, many geologists and astronomers proposed that an asteroid impact had caused the sudden extinction of the dinosaurs at the end of the Cretaceous period, sixty-five million years ago. Although strictly, dinosaurs lie outside our timeframe, the recent advances in geological thinking have broken down a number of barriers to that investigation.

Until 1980, the possibility of an asteroid impact at the Cretaceous-Tertiary (K-T) boundary was just one unproven theory among many. The breakthrough came when Luis and Walter Alvarez, and a team of researchers from the University of California set out to investigate the thin layer of clay that marks the K-T boundary in the rocks. Their results showed an unexpected abundance of the element iridium in the clay, together with other rare elements such as platinum and palladium.[33] Iridium in particular is rare in the Earth's crust, but is more common in meteorites. Also present in many places were grains of shock-metamorphosed quartz. Wherever in the world the layer was sampled, the results were similar. Alvarez asserted that this was firm evidence of an asteroid impact.

The original theory proposed that an asteroid between six and fourteen kilometres in diameter had hit the surface throwing up a dust cloud that enveloped the entire world for several years. In many places the plants died as they were denied sunlight and along with the plants went every creature in the food chain that depended upon them. The vegetation could regenerate from seeds, but the animal species, or most of them at least, were gone forever. Such hard evidence could no longer be denied and the search was on to find a sixty-five-million-year-old impact crater.

Geologists were aware of many structures that could be ancient craters, but none that fitted the K-T boundary event. However, this all changed in 1991 when a structure in the Chicxulub formation on the coast of Yucatan was identified as a buried crater of about the right age.[34] Indeed the 165-km diameter crater was found some years earlier during oil exploration, but the authors encountered great difficulty in getting their paper published. Some estimates suggest that as much as 80,000 km^3 of ejected matter was inserted into the atmosphere by the impact and the resulting dust-veil blocked out the light of the sun for several years.[35] This "impact-winter", according to current thinking, is what killed-off most of the higher life forms.

The Chicxulub crater was probably created by an encounter with a 10-km asteroid, or a somewhat larger comet. Astronomers have now calculated that such impacts by Earth-crossing asteroids and comets are the likely cause of most of the mass-extinction events identified by the palaeontologists. Similarly devastating impacts, are expected to occur about every 100 million years, with lesser, but still globally damaging strikes by bodies down to say: 1.5 km diameter occurring at intervals of about 70,000 years.[36] Arguably, these same conclusions could have been reached without the necessity to first discover a crater, but until the 'hard' evidence was found, no one was listening!

The debate as to whether there was a periodicity to comet and asteroid bombardment therefore rendered them at once a mechanism that is still present today. The realisation that the Earth could after all be bombarded from space renders it merely another geological agent. Charles Lyell's uniformitarian principle still holds true.

While the discovery of the Chicxulub crater has legitimised the debate about catastrophism on a *geological* time scale, perhaps surprisingly it has had only a negative effect upon the debate about the role of catastrophism in human prehistory. After all, why should we concern ourselves with a risk that occurs only at intervals of millions of years, and which probably will not occur again for millions more years yet? Furthermore, by revealing the terrible scale of such a disaster, it gives the obvious counter that if such a collision had occurred during human prehistory then humanity would be extinct. Alternatively, the crater would be large and obvious. An impact small enough to have left no trace could not deliver enough energy to cause either a pole shift or an axis tilt.

The academic debate therefore no longer centres on a worldwide cataclysmic event, but is limited to the local climatic effects of major volcanic eruptions; or just possibly the insertion of cometary dust into the atmosphere. As will be discussed in a later chapter, the evidence of dendrochronology and ice cores taken from the polar ice sheets reveal that there have certainly been some non-uniform climatic events during the past several thousand years. Scientists can safely debate these anomalies in the academic journals – so long as their language is suitably restrained.

The event that has really opened minds occurred on July 16 1994, when astronomers were treated to the spectacle of a comet striking the planet Jupiter. Here was present-day catastrophism viewed from a safe distance. Many astronomers, especially the comet's co-discoverers, David H. Levy, Carolyn and Dr Gene Shoemaker had never doubted the perils that exist in the universe. Levy was actually in the process of writing a book about comet hunting, when the remarkable event occurred – their discovery of a fragmented comet that was going to hit a planet.[37]

Comet P/Shoemaker-Levy 9 and its collision with Jupiter has given us all a warning of what would happen when a comet hits the Earth. More importantly, it has awoken everyone to the possibility that it can happen *now* – not merely in some safely remote geological epoch. Some astronomers had predicted a fizzle, a non-event.[38] However, when the first of the fragments struck Jupiter they were astonished to see a fireball 3,000 km high in the planet's atmosphere, which subsided into a vast dark spot, half the diameter of the Earth. Over subsequent days, the remaining twenty fragments left a string of black scars across Jupiter.[39]

Levy himself calculated that a 1 km size comet should hit Jupiter perhaps as often as once a century, compared with once in a hundred thousand years for the Earth. Astronomers searched the observation records for other dark spots reported by earlier astronomers. The only other comet known to have been disrupted by close encounter with Jupiter was Comet Brooks 2 in 1886, but it escaped. Astronomers now believe that Jupiter has protected the Earth throughout its existence, by sweeping the inner solar system clear of potential assailants.

Because of the millions who would die in such a rare collision, the odds of any individual person dying in a comet or asteroid strike are as low as 1 in 20,000. This would make it more likely than a lightning strike or a lottery win.[40] What Levy called the 'giggle factor' was gone. Astronomers and governments now openly assess the hazards and discuss how an asteroid might be diverted or destroyed. The Film industry has awakened the public interest by fictional impact scenarios, with Hollywood stars saving the world from a seemingly inevitable doom.

Where do these developments leave the subject of catastrophism in human prehistory? Certainly one no longer has to justify that it could have happened.[41] Rather, we should be able to assess the evidence, mythological as well as physical, without fear of the ridicule that has gone on in earlier decades. It should be possible to approach the subject armed with the latest discoveries, rather than leaving the field open for the wordsmiths of outmoded science.

The foregoing review is therefore not so much a review of catastrophism in human prehistory, for there are many other contributions that could have been mentioned. Rather, it is an overview of why the various theories have failed to gain acceptance. Moreover, the present author must show that despite the fact that he understands the orthodox science and chronology, and retains utmost respect for the painstaking work of the archaeologists, there nevertheless remains a case to be made for catastrophism.

If we probe the events of recent millennia within any single discipline, the evidence eventually becomes so tenuous that speculation begins to creep in and academic specialists dare go no further. It is necessary to bring together clues from various disciplines to construct a multidisciplinary pattern of evidence.[42] Looking at human prehistory from the viewpoint of mythology and astronomy and with the eye of a generalist, there is an anomaly, a dislocation in prehistory, which cannot be suitably explained within any single academic discipline.

In the first chapter, we reviewed the pattern of myths and legends that, from geographically distinct locations, gave a consistent pattern of a vaguely remembered flood or deluge; and the antediluvian world that it destroyed. In this chapter we have seen a few of the previous theories that have been put forward. We must now move on to review some of the evidence from other disciplines: calendars, chronologies and astronomy, as well as climate and sea level research. When we have assembled this pattern of clues to the past, will we be in a position to look again at the true nature of the Great Flood and its ultimate cause.

References to Chapter Two

1   Burnet, T (1681), The Sacred Theory of the Earth

2   Whiston W. (1696), A New Theory of the Earth

3   Steno, N (1669) reprinted 1968

4   Hooke, R (1705)
Lectures and discourses of earthquakes and subterraneous eruptions. My attention was first drawn to this because of his investigations of the 12,000-year time scale of Plato's Atlantis, which he considered far too long to be believable.

5   Werner (1786), A short Classification and description of Different Rocks

6   Hutton, J (1788), Theory of the Earth

7   Lyell, C. (1830-33), Principles of Geology

8   Buckland W. (1823), Reliquiae Diluvianae

9   Agassiz, L. (1840), Etudes sur les Glaciers

10 Darwin, C (1859), The Origin of Species

11 Jamieson, T. F. (1862) Quarterly Journal of the Geological Society, 18, 164.

12 Just as one example, see the article in Nature by Howard (1997).

13 Le Verrier, U (1855). This was the culmination of his work on planetary perturbations that had led to the discovery of the planet Neptune.

14 Croll, J (1870) Geological Magazine, 7, 271.

15 Hayes, J.D., Imbrie, J. and Shackleton, N.J. (1976)

16 See for example Nennius and his six ages of the world, or the similar ideas found in Geoffrey of Monmouth's Historia Regnum Britanniae.

17 Davies E. (1804), Celtic Researches and (1807) The Mythology and Rites of the British Druids.

18 For example: Champolion: 5867BC, Boch: 5702BC, Bunsen: 3623BC, Leiblein: 3893BC and Brugsch: 4400BC.

19 Professor Stuart Piggott, in a television program in 1971

20 See: Oeuvres Completes de Laplace (1835). Laplace in 1806 postulated that comets roamed interstellar space. For an English translation of the relevant parts, see Clube & Napier (1982).

21 Euler determined that the axis of rotation should describe one complete circle about the axis of figure in $A/(A-C)$ days, where C is the moment of inertia about the polar axis and A about an equatorial axis. The mathematics of the Earth's rotation would never seem as simple again as it did in 1765.

22 See Howarth, 1871 & 1872.

23 It is interesting to follow the evolution of Jeffreys' arguments through the twentieth century, in the revised editions of his book *The Earth*.

24 See Bullard, E.C. (1975). However, the North-South polarity does reverse at intervals of half-a-million years or so. The transitional period is too rapid to be recorded in the rocks. It was the record of these reversals parallel to the mid-ocean ridges that finally proved the continental drift theory in the 1960s.

25 I first encountered these ideas when I was about ten years old, in a piece of children's fiction entitled Space Agent and the Ancient Peril (Angus

MacVicar, Burke, London 1964) and I recall that even then I had enough knowledge of astronomy to dismiss it. If anyone should wish to pursue the bibliography it may be found in the following books:

H. S. Bellamy, Built Before the Flood, second edition (1947)

Moons, Myths and Men, second Edition (1949) The Calendar of Tiahuanaco (1960)

A. Posnansky, Tihuanacu, the Cradle of American Man (1945)

E. Kiss, Das Sonnentor von Tihuanaku (1937)

H. Hoerbiger and PH. Fauth, Glazialkosmogonie (1913).

26 Joshua 10, 12-14. Such a standstill of the Sun is difficult to rationalise by any known physics, therefore we have to resort to the unknown. Perhaps this marks the passage of a gravity wave through the Earth, or an encounter with the so-called dark matter.

27 The foreword to Hapgood's book was written by no less a person than Albert Einstein, which certainly added to its credibility.

28 This is because a book with pyramids or 'civilisation' in the title will generally end up in the Egyptology or Ancient History section of the library or bookshop, where it is more likely to be noticed. A book with 'stars' or 'planets' in the title will end up under astronomy. However, a book about Atlantis will usually go in the miscellaneous section. There is no recognised category to market a book about mythology, folklore or catastrophism. I have found my own book The Atlantis Researches: the Earth's Rotation in Mythology and Prehistory under all three categories. I have found Peter James's scholarly work The Sunken Kingdom: The Atlantis Mystery Solved filed under miscellaneous, while Graham Hancock's Fingerprints of the Gods was in the same library under Egyptology. What's in a name? Everything it would seem!

29 Perhaps the least credible of these books are those that pursue the predictions of the psychic Edgar Cayce, which have a pedigree similar to Mme Blavatsky's Theosophy. Cayce predicted in 1940 that part of Atlantis would rise again from the sea in 1968 or 1969. When in 1968 an American zoologist discovered what appeared to be a submerged harbour or jetty off the Bahamian island of Bimini, the so-called 'Bimini Road' was immediately hailed as Cayce's prediction coming true! The host of 'new age' books that have built upon this base have done enormous damage to the credibility of catastrophist enquiry in general. Why so many people prefer occult predictions to science must remain a mystery in its own right.

30 When I first heard this theory I was reminded of the floating island of San Serif, an April Fool joke once perpetrated in a British newspaper, and which was supposedly floating up the North Atlantic to collide with Britain! Most journalists recognised Sans Serif as the typeface used by many newspapers.

31 Collins himself was another advocate of a Caribbean cataclysm in his book Gateway to Atlantis

32 For a detailed critique of Hancock and Bauval and various other 'alternative Egypt' theorists. then see Giza: the Truth by Ian Lawton and Chris Ogilvie-Herald.

33 Alvarez, L.W. et al, 1980 and 1982.

34 Hildebrand, A.R. et al., 1991.

35 Morgan, J, et al, 1997, p 476.

36 Chapman, C and Morrison, D, 1994, pp 34-35.

37 Levy, D.H. The Quest for Comets, 1994.

38 Weissman, 1994, in an article in Nature.

39 The impact velocity of the Shoemaker-Levy-9 fragments was estimated at 60km/sec, with each fragment of the order of 100 times smaller than the KT impactor. See Melosh, 1995.

40 Clark Chapman, at a conference in St Petersburg in 1991.

41 For example, in my earlier book The Atlantis Researches, written in the early 1990s and published in 1995, I had devoted an entire chapter to justifying the possibility of a comet impact. Such justification is simply no longer necessary.

42 For example, in a millennium essay in the magazine Nature, F Seltz, bemoaned "the decline of the generalist"; see millennium essay, Nature, p 403, p 483, 3 February (2000). Generalists are alive and well. The problem lies in the availability of avenues to publish their work.

3

Sun, Moon and Calendar

If we are to make sense of the many references to astronomy in ancient myths, then we have no alternative but to grapple with the evidence that has come down to us in references to the calendars of ancient people.

Most of us are taught the basics of the calendar as part of our earliest education. We never need to observe the sky to work out the rules for ourselves; all of this work was done for us millennia ago by our ancestors. We are taught that the year has three hundred and sixty five days and twelve months, each with thirty or thirty-one days, "save February alone which has twenty eight", but these 'months' have little to do with the real Moon. We all know that because the true year is not an exact number of days then every fourth year we must add a day to February and that this is a 'leap-year'. A few may know that there is also a century rule to ensure that the calendar year remains even more closely in step with the seasons. Christians may know that the date of Easter, unlike Christmas, falls on a different date each year and that this has something to do with the Moon. Most people's interest in the calendar ends here.

The Gregorian calendar, now almost universally adopted, is but a refinement of the calendar reforms introduced throughout the Roman Empire by Julius Caesar in 46BC to correct the errors that had accumulated in the old Roman calendar. Even so, the Julian calendar was allowed to drift out of season by ten days before Pope Gregory XIII in 1582 dared to introduce the century rule and omit the excess days from the Calendar. Other nations such as Britain and her American colonies did not adopt the reform until 1752; and Russia not until 1918. This shows us how tolerant people can be of calendar inaccuracies, if they have been brought-up with them.

The names of the months are even older than Caesar and such names as Sept-ember and Oct-ober tell us (if we did not know) that the old

Roman year formerly began at the vernal equinox rather than midwinter. Other month names preserve for us the names of pagan gods of Rome – despite the adoption of the Roman calendar by the Christian church. Alongside the month runs the seven-day cycle that we call the week. In English this is even more revealing, because the names of some days recall pagan Germanic gods such as Tuesday for Tiw and Thursday for Thor; other names reveal the astrological origins of the calendar: the Sun's-day; the Moon's-day and Saturday for Saturn. The Julio-Gregorian calendar illustrates for us that calendar traditions, wherever we may encounter them, may have roots that date back many millennia. Calendar tradition is incredibly conservative.

The calendar that Julius Caesar introduced is a solar calendar, but it replaced a lunar calendar. In a solar calendar, the cycle is held in step with the seasons and the cycle of the Moon is allowed to drift freely. Other cultures, including most of the Muslim world, prefer strict lunar calendars. These instead repeat the lunar month, but again since the orbit of the Moon is approximately twenty-nine and a half days, a sequence of alternating twenty-nine and thirty-day months must be employed. The fundamental problem underlying all calendars is that the cycles of the Sun and the Moon are not commensurate. A lunar year of twelve months (12 x 29°) adds up to only 354 days, about eleven and a quarter days short of the solar year. The ideal is therefore to devise a *luni-solar calendar*, in which *intercalary months* are inserted over a period of years. The two cycles then come back naturally into step and will repeat indefinitely.

Alas there is no such ideal period. The ancient Greeks tried inserting three extra months every eight years – a cycle known as the *octaeteris*. A cycle with four intercalary months in eleven years would be closer; but one with seven intercalary months over nineteen years is better still. (see table 3.1) This is known as the Cycle of Meton after the Greek astronomer who introduced it from Babylon. The same cycle is still used in the modern Jewish calendar.

Other intercalation cycles could be employed to make a perfectly acceptable luni-solar calendar; the Calippic cycle of 76-years (4 x 19) is better still and an 84-year cycle is nearly as good.[1] The crucial consideration is how many years elapse before the calendar requires further arbitrary intervention. The Romans pontifices were hopeless at this intercalation and by the time of Caesar's drastic reform their festivals had drifted three months in advance of the proper season.

Table 3.1 The accuracy of various intercalation cycles

Inter-calation cycle (y)	Inter-calated months	Synodic month	No. of months (x)	Days in x months	Tropical year	Days in y years	Discre-pancy (days)
3	1	29.530589	37	1092.63179	365.2422	1095.7266	3.094807
3	1	29.530589	37	1092.6318	365.2422	1095.7266	3.094807
8	3	29.530589	99	2923.5283	365.2422	2921.9376	−1.590711
11	4	29.530589	136	4016.1601	365.2422	4017.6642	1.504096
19	7	29.530589	235	6939.6884	365.2422	6939.6018	−0.086615
22	8	29.530589	272	8032.3202	365.2422	8035.3284	3.008192
30	11	29.530589	371	10955.849	365.2422	10957.266	1.417481
57	21	29.530589	705	20819.065	365.2422	20818.805	−0.259845
76	28	29.530589	940	27758.754	365.2422	27758.407	−0.34646
84	31	29.530589	1039	30682.282	365.2422	30680.345	−1.937171

This conservatism in calendrical tradition will go some way to explaining why most people will react with horror to the present author's suggestion that the Earth's axis and the length of the day may have changed in ancient times. It throws our most basic assumptions about the world into confusion. However, if the day, our fundamental unit of time, has indeed changed then it should reveal itself in the world's oldest calendar traditions.

The earliest civilisations seem overwhelmingly to have preferred lunar calendars to solar. We may ask why this was. The conventional view is simply that our early ancestors were not competent enough at astronomy and mathematics to work out the realities of the world. This, as we shall see, is far from being borne out by the evidence.

Astronomy in Ancient Babylon

From a very early era, the Sumerians divided the yearly path of the Sun into twelve equal sections (roughly 30° for each sign of the zodiac). This was used to reckon the motion of the Sun and Moon, against the background of stars, over the course of a year. Thus the year could be regulated by 12 x 30 = 360 parts. These evolved into the twelve constellations of the zodiac, which are still the basis of all modern astrology. A Babylonian astrological text survives from about 700 BC, which describes the regulation of the year by a stellar calendar of 360 days. However, there are other indications that the zodiac itself may date from as early as 3500 BC.[2]

The Sumerians also used a lunar cycle for civil purposes, which was regulated by direct observation of the first crescent. Records of the insertion of intercalary months survive from as early as 2400 BC. The solar year was determined by observing the seasonal rising of the prominent stars in each zodiacal sign. If the rising of a marker star for that month's constellation had drifted a calendar month late then the priests would decree the insertion of an intercalary month, making that year one of thirteen months (or sometimes even fourteen months if the priests had been lax). Thus were Sun and Moon kept in step.

The Sumerian practice was continued by the Old Babylonians. An inscription from the reign of King Hammurabi (c1700BC) declared that the payment of taxes must not be delayed by the insertion of an intercalary month! The calendar was subsequently adopted by the Assyrians (c1100 BC), and by conquered peoples, such as the Jews. After the Persian conquest in 539 BC, it became the official calendar throughout their empire, from India to Asia Minor.

It was not until as late as the sixth century BC that the insertions of the intercalary months were made to follow a regular cycle of seven intercalary months over a period of nineteen solar years. This was the cycle that Meton displayed to the Greeks in 432 BC. The necessity to regulate the calendar by direct observation, for civil and astrological purposes, may therefore explain why the Babylonians preserved so many cuneiform records of their astronomical observations.

The high standard attained by the early Babylonian astronomers can scarcely be doubted. When Alexander the Great conquered Babylon, his astronomer sent back to Aristotle a list of eclipses dating back nearly two thousand years. Classical writers such as Geminus, Ptolemy and Pliny all held Babylonian astronomy in the very highest esteem; Pliny for example cites Epigenes:

> ...an authority of the first rank, teaches that the Babylonians had astronomical observations for 730,000 years inscribed on baked bricks; and those who give the shortest period, Berossus and Critodemus, make it 490,000 years.

Ptolemy, in his *Almagest* similarly tells us that records of lunar and planetary phenomena were systematically preserved by the Chaldaean astronomers from the reign of the Assyrian king Nabonassar (747 BC) up to his own time. The very word Chaldaean would henceforth come to mean 'astronomer'. We are told that Nabonassar collected and destroyed the records of earlier kings, so that the reckoning of future kings would begin at his reign – the Era of Nabonassar.

Archaeological evidence also supports the pre-eminence of Babylonian astronomy. Many such astrological tablets were discovered in the library of King Ashurbanipal, the same collection that held the Gilgamesh tablets. Observations of Venus and the Moon are known, which date back to the early second millennium BC, although their precise date remains problematic.[3] We may ask why the Babylonians did not discover the nineteen-year cycle or some equally suitable intercalation rule much earlier, if they were indeed such capable astronomers.

Following Alexander's conquest of the Persian Empire, many more Babylonian ideas found their way back to Greece. Our modern system of time measured in hours of sixty minutes and sixty seconds is based upon the Babylonian system, casually adopted by the Greeks and subsequently by the Romans. The same notation was used for divisions of the circle and we still use it today. For the Babylonians counted in units of sixty. From at least as early as 1800BC and certainly by the Seleucid period (after 323BC) the same system was used in astronomical calculations.

Just why the Old Babylonians chose this sexagesimal system remains uncertain, but it probably had its origins in their astronomy. In cuneiform writing, they would repeat a simple T-shaped wedge for numbers from 1-9 and a caret-shaped wedge for 10. They would repeat this six times to make sixty. For larger numbers they would show multiples of sixty, plus the units; and for even larger numbers they employed a unit of 'sixty squared' or 3600, which Berossus tells us was called a *saros*. As they had no symbol for zero, the absence of a number in any place had to be represented by a space, which could easily lead to error for both ancient and modern readers alike.

Berossus records other intervals: the *neros* of 600 years and the *sossos* of 36 years. This has led to the, wholly credible, suggestion that the division of the circle into 360 'degrees' was a consequence of sexagesimal arithmetic, representing a tenth part of the saros, or ten times the sossos; and that this was used as a best approximation of a solar year.

Calendars and chronologies are interrelated. In order to specify a calendar date, the years and months must be counted from a fixed *era*. Usually this date will have some meaning within the religion or culture. For example, Gregorian dates are numbered from the birth of Christ, Muslim years from the flight of the Prophet and so forth. In the earliest period for which we have evidence, the Babylonians seem not to have needed an era. They numbered their years from the accession of each king, which is adequate for short-term civil purposes. This may explain why they found it so important to maintain king lists, as a means of counting back dates into history.

Berossus and the Flood

Perhaps we may find clues to the earliest practice in the fragments of Berossus. When Alexander the Great and his Macedonian armies overwhelmed the former Persian Empire, they gained hegemony, in Egypt, Babylon and India, over cultures far older than their own. Although snippets of Babylonian astronomy and Egyptian history found their way to the west, the Greeks themselves could not read the cuneiform and hieroglyphic texts. So around 281 BC, Berossus, a Chaldaean priest, attempted to summarise the chronology and mythical history of his native country in a Greek book called the *Babyloniaca*, dedicated to his Seleucid-Greek ruler Antiochus I. Its importance as a source of the Babylonian Flood tradition has been discussed above.

The Babyloniaca was influential to many Greek and Roman authors. The importance of Berossus for the present enquiry lies in his use of the king lists and mythological sources, together with the Babylonian cycles of time, to reconstruct the date of the Flood. It is interesting that he alone among our sources gives us a precise calendar date. He says that it began on the fifteenth day of the Babylonian month of Daisios (= May) which was the second month of the Babylonian year. This simple statement, if taken at face value, would indicate that a continuous calendar tradition had been preserved in Mesopotamia since the Flood; *and that a calendar of some kind had existed even before it*. So what might the antediluvian calendar of Mesopotamia have looked like?

The Biblical account gives us some quite precise data on the duration of the Flood, from which we may extract useful information. Here we are told that the Flood began on the seventeenth day of the second month, in the six hundredth year of Noah's life.

<u>Days elapsed</u>

Month 2 day 17: Noah enters the ark	0
Seven days later the Flood begins	7
The rain falls for forty days	47
The waters began to recede after 150 days	150
Month 7 day 17: the ark rests upon mount Ararat	c.150
Month 10 day 1: mountains were seen.	c.224
Forty days later a raven is released	264
Seven days later a dove is sent, but it returns	271
Seven days more, a dove is released and returns with an olive	278
Seven days more, a dove is released and does not return	285
Day 1 of Month 1 of Noah's 601st year, the ark is opened	c.284
Day 27 of the second month: the earth is dry again.	c.370

We may see that a year of Noah's life has passed and that the total duration of the voyage was a year and ten days. If we assume that each month has 30 days then the total of all the days mentioned comes to: (12 x 30) + 10 = 370 days, give or take a day if the months were variable. The day count does not add-up if we assume a lunar year of 354 days, nor if we use a 365-day solar year.[4] It seems to imply a year of 360 days.[5] This tells us with some certainty that the details of the Flood were not preserved according to the later Jewish or Babylonian lunar calendar, yet the six hundred years [= *one neros*] of Noah's lifespan serve to confirm its ultimate Chaldaean origins.

The association of a 360-day year with the Flood is also evident in the apocryphal Book of Enoch. It asserts that nature and the heavenly bodies were perverted owing to the sinfulness of mankind. Enoch laments that in the days of the sinners (i.e. before the Flood) the years were "shortened" and that "all things on the earth shall alter".[6] The calendar year is described as a solar year of 364 days, with four intercalated days at each quarter of the year. These four extra days, he says, were not counted in the original reckoning of the year.

There is no way to date the origin of these traditions. However, it is clear that they record an early religious belief that the Flood brought about a real change to the number of days in the year, and does not merely describe a reform of the calendar.

Calendar Tradition in Ancient Egypt

According to Herodotus, the Egyptians considered themselves the oldest civilisation in the world; they believed they were the first to divide the year into twelve parts and to devise a year of 360 days.[7] Other classical writers too, remarked on the quality of Egyptian astronomy. Diodorus of Sicily went further, saying that the Egyptians preserved records of the stars over an incredible number of years, and that they employed them to foretell many things, saying:

> ... and as a result of their long observations they have prior knowledge of earthquakes and floods, of the risings of the comets, and of all things which the ordinary man looks upon as beyond all finding out. And according to them the Chaldaeans of Babylon, being colonists from Egypt, enjoy the fame which they have for their astrology because they learned that science from the priests of Egypt.[8]

Most Egyptologists would hold that there is little evidence to support the claims of Diodorus for an advanced Egyptian astronomy. Nevertheless, we should never dismiss native tradition, or the opinion of historians who were closer to the events than ourselves.

The civil calendar in use during the classical era was a solar calendar based upon 12 months of 30 days. At the end of each year the Egyptians added five extra days, called epagominal days by the Greeks, to complete the year of 365 days.[9] As this was still a quarter day short of the true orbit, the calendar fell behind the seasons by a full day every four years, taking 1461 (Egyptian) years to come back into step. There is no evidence that the Egyptians ever adjusted the civil calendar to prevent this cumulative error.[10] The civil year was favoured by astronomers such as Ptolemy, precisely *because* it was never adjusted. Unlike a lunar calendar, they did not have to guess in which years an intercalary month had been inserted.

Egyptologists call this long cycle of the civil calendar the *Sothic Cycle*, after the Graeco-Egyptian name for the star Sirius. In a manner very similar to the Chaldaean zodiacal star calendar, the Egyptians used the pre-dawn, or heliacal, risings of stars as seasonal indicators. This is evident from a very early inscription, dating from the First Dynasty (3100-2900BC) which refers to Sirius as the marker of the Nile flood season.[11] Every year, around mid July the star reappears in the morning sky, having been obscured by the sun for some seventy days.

However, the earliest evidence, dating from the fourth dynasty (mid-third millennium BC) describes a lunar calendar. Egyptologists have painstakingly reconstructed its workings. The start of the lunar year was constrained to coincide with the pre-dawn rising of Sirius and thereby with the annual flood of the Nile. An intercalary month named after the Moon god Thoth, was thus inserted whenever the heliacal rising of Sirius fell within the last eleven days of the twelfth month.

As the first month of the civil calendar was similarly named after Thoth, Egyptologists therefore presume that it was devised at an era when the first day of the civil calendar coincided with the first day of the lunar calendar.[12] Although this may be a somewhat speculative assumption, it nevertheless formed the basis of Egyptian chronology for many years.

The way into the chronology was supplied by the Roman author Censorinus, who recorded that 20 July 139 on the Julian Calendar corresponded to day one of Thoth. This simple statement allowed Egyptologists to retrocalculate that the first day of the civil calendar coincided with the pre-dawn rising of Sirius in 1322 BC and earlier in 2782 BC and 4241 BC; but because of Manetho's king list, most authorities now prefer 2782 BC for the origin of the civil calendar. Dates associated with various kings within inscriptions have enabled Egyptologists to further refine the chronology.

The five epagominal days were named for the five principal gods of Egypt and are found as early as an inscription in the Fifth Dynasty *Pyramid Texts* dating from c2400 BC and which continued to be carved within the pyramids of Sixth Dynasty kings. A text on ascension and rebirth, states:

> The prince ascends in a great storm from the inner horizon;
> he sees the preparation of the festival, the making of the braziers,
> the birth of the gods before you in the five epagominal days...[13]

This would seem to suggest the establishment of a 365-day year earlier than the Fifth Dynasty.

According to Manetho's chronicle, during the reign of a Fifteenth Dynasty king, (c.1600 BC) twelve hours were added to the month to make it 30 days long and six days added to the year. Precisely what this refers to, nobody knows. Clearly, it describes a calendar reform of some kind, but is not otherwise confirmed. If it recalls the addition of six days to a 354-day lunar year, then it implies a 360-day year. If it signifies the addition of the epact to a 360-day year then it would imply an *earlier* 360-day calendar. A *sixth* epagominal day would be needed only every fourth year, but as we have seen, Egyptologists find no evidence that the calendar was ever fixed.

This begs the question of *why* a year of 360 days was ever chosen at all? Here again, mythology may help us. The Greek author Plutarch, in his moral essay *Isis and Osiris*, has preserved a calendar myth explaining how the epagominal days came to be named after the five principal gods. The *Myth of Osiris* states that Rhea, wife of the Sun god had an affair with Cronus.[14]

> They say that the Sun, when he became aware of Rhea's intercourse with Cronus, invoked a curse upon her that she should not give birth to a child in any month or any year; but Hermes [=Thoth], being enamoured by the goddess, consorted with her. Later playing draughts with the moon, he won from her the seventieth part of each of her periods of illumination, and from all the winnings he composed five days and intercalated them as an addition to the three hundred and sixty days. The Egyptians even now call these five days intercalated and celebrate them as the birthdays of the gods.[15]

Therefore, we may see in this myth, not so much a change in the calendar, rather a belief that in the earliest times the true year had been 360 days long; *or rather that the period of a year formerly contained only 360 days*. This is further suggested by the statement that the Moon's periods of illumination (i.e. the nights) were shortened by a seventieth part.[16]

So again, if we take mythology at its word rather than dismissing it: does the Myth of Osiris record that at some period in the past, the length of the day decreased, such that 365 days made a year? Does it recall that in an earlier epoch the year contained only 360 elapsed days? It would certainly explain why the constellations were chosen to divide the zodiac into 360 parts rather than 365. Could the cause of such a drastic change to the basic parameters of our world be none other than the Flood, the great Deluge of which the myths tell us? If so, then we should expect to find similar devices world-wide in the calendars of other ancient civilisations – and indeed we do!

The 360-day Calendars of India

Shortly after the British colonial administration departed from India, the new government instituted a survey of the calendars in use throughout the country. It revealed that no less than thirty different calendars and eras were used by the various religious sects.

It is evident that many of the workings of the Hindu calendars were already well-established by the earliest period of the sacred texts known as the *Vedas*. These epics are believed to have been composed around 1500 BC when Aryan invaders from central Asia overran the earlier Harappan civilisations of the Indus Valley. As the Harappan script has never been deciphered, we know nothing of their mythology or astronomy.

However, most Indian astronomy can be traced with certainty, only as far back as the astronomical texts known as the *siddhantas* or 'solutions', of which the earliest text is no older than the first century AD. All these sources reflect the Babylonian influence that reached India with the conquest of Darius I in 519 BC.

The Indian solar calendar functions superficially like the Julio-Gregorian calendar, in that it has twelve 'solar months', varying between 29 and 32 days each. However, while users of the Western calendar need take little interest in the real sky, Indian practice ties the length of each month closely to the date that the sun actually enters each zodiacal sign. The constellations are known as *sankranti* and the year begins at Mesha-sankranti, which is the equivalent of Aries.

Astronomical references within the Vedas reveal an early luni-solar calendar with 12 months of 30 days each and a year of 360 days.[17] The details of this Vedic calendar are uncertain, but it appears that the years were arranged in a five–year period called a *yuga*. Rather than adding five days at the end of each year, as the Egyptians did, later Indian custom has been to store-up the discrepancy until a full extra month could be inserted.

India's calendars may also preserve something of the way that the marker stars were used in ancient Babylon. Each of the 27 constellations contains a marker star or *nakshatra*. These are chosen such that each is about 13° 20′ apart. The Moon therefore passes close to each marker during its sidereal period of 27.3 days [*360 ÷ 27 = 13.33*]. Thus, the Moon is regulated by its position in the sky rather than by the lunar phase. In the south, months began at the first crescent after new moon; in the north, they began after full moon.

Indian calendars not only intercalate an extra month, but may also omit a month. Usually the Moon reaches its conjunction (either full or new moon) only once between any two marker-stars, however, if two conjunctions occur in the period of a sankranti then an intercalary month of the same name is repeated.[18] Less often, two sankranti may fall within a lunar month, in which case the month following is omitted. This mechanism is self-correcting, the average of these machinations will ensure that there are seven intercalations in nineteen years.

Therefore, it is not possible to state with certainty how much Babylonian astronomy has influenced India and how much India influenced Babylon, still less any early exchange with Egypt. Although the practice of dividing the sky into 360 parts may be very old, the astronomy is certainly *not* primitive. Indian calendar makers required an extensive knowledge of astronomy. This will be become still more evident as we go on to examine, in the next chapter, the eras and cycles that they devised.

The Sixty-Day Cycles of Ancient China

Most people will know of the Chinese "animal" calendar, even if they know little else about China. Each Chinese year is named according to a twelve-year cycle of animal names, or 'earthly branches'; for example 2000 was a year of the dragon, 2001 a year of the snake and so on. This twelve-year cycle is also meshed with a ten-year cycle of 'heavenly stems' to make a repeating sixty-year cycle. Every year in the sixty-year cycle can thus be uniquely named. The same system is followed throughout south-east Asia and Japan.

This sexagesimal system is very ancient and is thought to date from the third millennium BC. The earliest direct evidence comes from the so-called 'oracle bones' dating from the Shang Dynasty (after 1600 BC) which equates to the Chinese Bronze Age. In the 1930s a horde of these divinations were discovered, on bones and turtle shells, near Anyang on the Yellow River.

The oracle bones show that the symbols for the heavenly and earthly branches were originally used as day counts in a system of calendar dates. From the Zhou dynasty (after 1000 BC) the same symbols were also given to the years, after which the day count seems to have lapsed into disuse. However, it may be seen that, in the earliest period, five of the sixty-day cycles made a year of three hundred and sixty days. By its very nature, there is no room in such a scheme for a corrective system of epagominal days and the sixty-day cycle therefore repeated indefinitely.[19]

The sexagenary cycle ran alongside a lunisolar calendar, which is also found on the oracle bones. It seems to have worked, superficially at least, in a similar way to the Indian lunisolar calendar. The ecliptic was divided into twelve solar months, analogous to the Indian sankranti and the western zodiac. However, the Chinese used 28 marker stars, called *hsiu* or 'lunar mansions', rather than 27 as in India, and many of the marker stars are different. Whenever a new lunar month began within the same hsiu, it was adopted as an intercalary month. As with the sixty-day cycles, there is no way to know if the concept of 'solar months' was borrowed from Babylon or India; or indeed, whether they each invented it independently.

Figure 3.1
A Shang oracle bone, with an inscription believed to refer to an eclipse.

Chinese astronomers of the Han Dynasty (260 BC to AD 221) also knew the nineteen year Metonic cycle, which they called the *chang*; and the seventy-six year cycle or *pu*. Oracle bone divinations concerning eclipses would indicate that these resonances of the Sun and Moon were known at least as early as the Shang period (c1300 BC), far earlier than any comparable western evidence.

The Chinese calendar has therefore evolved by increasingly accurate refinement of the measured lunar month and solar year. Measurement of the year is evident in the earliest historical records in *Shu Ching*, the Chinese historical epic. During the reign of the semi-mythical emperor Yao (c2100 BC) we find the Chinese establishing the true period of the year by measuring the length of the shadow, usually at the winter solstice.

Ever increasing accuracy was achieved by using taller and taller gnomons to cast the shadow. It is also likely that the lunar month was refined by observation of eclipses, from at least as early as the Shang dynasty.[20] This gives the lie to any suggestion that an early belief in a 360-day divinatory year was somehow due to poor or primitive astronomy.[21]

Perhaps more interesting is Chinese treatment of the circle. From the earliest known records, Chinese mathematicians have always divided the circle into 365¼ parts, which clearly shows that its origin was based on the days of the year. This is what we should expect if their calendar depended upon direct measurement of the year from solstice to solstice.[22] They never used sexagesimal arithmetic as the Babylonians did.

In *Shu Ching* there are a number of astronomical references that have enabled historians to attach a date to some of the events. *The Canon of Yao* describes the mythical brothers Hsi and Ho who regulate the four cardinal points of the equinoxes and solstices. Each point is associated with a marker star and it may be seen that these do not correspond to the stars that currently mark the Sun's position at these seasons. Various scholars have therefore used the phenomenon of the *precession of the equinoxes* to back-calculate when the stars would have occupied these positions.[23] The historian J. B. Biot showed that this corresponded to the situation around 2400 BC.[24] At this era we are told:

> ...a round year consists of three hundred, sixty and six days. By means of the intercalary month, you are to fix the four seasons and complete the determination of the period of the year.[25]

Most scholars assert that the Shu Ching is a later forgery or revision, dating from the late centuries BC. Yet such a forgery could not have been perpetrated without an awareness of the precession. As far as we know, the Chinese did not discover the precession of the equinoxes until the fourth century AD.[26] This lends an authenticity to the passage. The reference to a year of 366 days at such an early era is problematic. It may imply the extra leap-day every fourth year – yet the passage itself implies a lunar calendar with intercalated months, which requires no such epact? The context implies that it was a measured year derived from the length of the shadow.

Thus, in Chinese calendrical science, we see a system that has been continuously updated by astronomical measurements, quite unlike the practice in India, where religious conservatism has tended to preserve ancient methods.

Central America and the Mayan Calendar

Orthodox archaeology would hold that the civilisations of the New World developed completely in isolation from Europe and Asia. The Maya civilisation in Central America reached its height only around AD 800, but may be related to the earlier Olmec and Zapotec peoples, dating back to around 1000 BC. As the Spanish conquistadors destroyed most of the Mayan religious texts, only a few books and inscriptions have survived to tell us of Mayan science. However, other cultures of Central America, such as the Aztecs, all used a very similar calendar, which was certainly known to the Olmecs before 100 BC.

The Maya of the classical period used a 365-day year known as the *haab*. This was split into 18 periods called *uinals*, each of 20 days – for the Maya counted in units of twenty. As with Egyptian practice, this made a year of 360 days to which an epact of five days was added. The year was allowed to wander through the seasons without correction in a cycle of 1460 years – again similar to the Egyptian Sothic cycle.

Alongside the haab ran another cycle of 260 days called the *tzolkin*, which was further divided into 13 periods of 20 days. Thus, it was possible to name any day uniquely by its place in either of these cycles, which would repeat only every 52 years:

$$52 \times 365 = 73 \times 260 = 18,980 \text{ days}$$

The significance of this 260-day sacred almanac is not documented, but a number of theories have been proposed.

The Maya also devised a system of dates called the *Long Count*, which they used to specify historical and mythological dates from a remote era. Discussion of this era may wait until the next chapter, but suffice to say that it parallels the dates AD and BC in the Gregorian calendar, with one important exception. Before the start of their era, they counted back only in years of 360 days, known as the *tun*. Larger multiples of day counts were made in multiples of 20, for example: 1 *katun* = 20 × 360 = 7200 days; 1 *baktun* = 20 × 7200 = 144000 days; and so on.

However, we may also note that:

$$52 \times 360 = 72 \times 260 = 18720$$

The significance of the number 72 has been discussed above in connection with the Egyptian Myth of Osiris. We may see that:

52 x 5 = 260 and
(52 x 360) + 260 = (52 x 365)

Therefore, it is the present author's opinion that the tzolkin cycle has more to do with the importance of the numbers 72 and 73 than with 260.[27] It enables 360-day years to be more easily mapped to 365-day years. There is no need to look for any more mystical purpose behind the tzolkin.

None of this explains why a 360-day year was chosen in the first instance. It may be seen that the calendar is loosely solar and took no account of the Moon. Alongside the long-count dates in many Maya inscriptions, there would be separate glyphs for the Moon and its phase; so unlike the other cultures discussed, it cannot be argued that the 360-day year was some loose attempt at a lunisolar year. The Maya were obsessed with the cyclical nature of time and the timing of sacrificial rites. As we shall see, they were well capable of calculating the appearances of the Moon and the planets.

Although archaeologists would like to consider the Central American calendars as no older than the first evidence, we should recall the example of the Julio–Gregorian calendar made at the start of this chapter. Calendars, wherever in the world we encounter them, have *always* evolved from ancient roots.

Lost Civilisations and Ancient Voyages?

How then do we account for the prevalence of 360-day years, in the calendars of all the earliest civilisations? Firstly, we should appreciate that it is really, not difficult, to establish the true length of the year; next to counting the days between the phases of the Moon, it is probably the easiest of all astronomical measurements. We have seen that the Chinese measured the year by the length of a shadow cast by a simple gnomon; and probably the same was achieved with Egyptian obelisks and the standing stones of Neolithic Europe. One might accept an accuracy of 365 or 366 days, or even 364 days, but not a measurement error of 5¼ days.

The other possibility is that 360 was chosen because of the properties of the number itself, it being a multiple of calendrically important numbers: 5 x 72, 18 x 20, 12 x 30 and 6 x 60. While 360 may be convenient for both vigesimal and sexagesimal arithmetic, how can we be sure which developed first: the astronomy or the arithmetic?

While we might accept a diffusion of calendrical science from Babylon to India, Egypt and China, it is hard to accept that the civilisations of America are anything but indigenous. Some authors would like to see astronomical wisdom preserved from a 'lost civilisation' dating back to the Ice Age, or carried across the Atlantic by ancient mariners from Egypt. Yet, any resemblance between the Mayan and the Egyptian calendar is wholly superficial. There is no need to resort to any kind of diffusion. The Sun, Moon and stars are the same wherever in the world they are observed. That ancient astronomers, all over the world, should have devised very similar systems is testimony to their competence rather than the reverse.

The fact remains that we always find the belief in a 360-day year in the very *oldest* evidence and in mythological sources that take us back to the time of the Flood. The evidence so far discussed suggests that we must go back to the late fourth millennium BC, in order to approach the date of the dislocation in prehistory that gave rise to these beliefs. Therefore, when we see evidence passed down from ancient astronomers, which tells us that the year formerly had only 360 days in it, then perhaps we should take a little more notice.

The calendar evidence may be giving us a clue that around the time of our earliest historical records, the length of the day was changed. We should be in no doubt that this was a catastrophic event, with consequences for world climate and sea levels that we find in the various Flood myths. The people who lived through this time would have inherited a calendar tradition that the year was 360-days long. Very soon, these survivors would discover that their calendar no longer functioned. The astronomers among them would have to adapt the calendar to the new reality.

However, the myths also tell of other changes to the Earth. A Chinese myth, for example, indicates that the axis tilted. The myths of devastating sea floods and emergence of islands suggest a pole tide associated with a wobble or nutation. These phenomena would also need to be reflected in the calendar and we should expect to find some indication of these cycles too, in ancient astronomy. In the next chapter, we shall go on to look for evidence of these effects in the eras and cycles devised by the earliest civilisations.

References to Chapter Three

1   The 84-year cycle is known to have been used by the early Christians and by the Picts of Scotland.

2   There are good reasons to believe that the Pleiades star cluster in Taurus, known as 'the bull of heaven', marked the point of the spring equinox at the era when the zodiac was first devised. Due to the precession of the equinoxes, the equinox has now precessed through Aries into Pisces, but would have been in Taurus between 5000-6000 years ago. On this see W. Hartner (1965).

3   See below, Chapter Six.

4   See A. Heidel, 1971, pp245-248.

5   The Jewish Pharisees used a 360-day 'lunar' year, with arbitrary intercalation, before the adoption of the Babylonian lunar calendar during their exile in Babylon.

6   Enoch LXXX, 2-8

7   Herodotus, 2, 2-7

8   Diodorus Siculus, 1,81,4-7

9   Literally "(days) brought in", or the excess of the solar year over the lunar year.

10   The Romans termed the Egyptian year the vague or wandering year. When a calendar reform was proposed in 238 BC by the so-called Decree of Canopus it was blocked by the priests. The decree records that winter festivals were occurring in summer and vice- versa. Presumably, the priests knew of some arcane reason that barred any interference with the calendar.

11   On an ivory label discovered by Sir Flinders Petrie at Abydos. See Parker 1950, p 34. A similar reference may be found in Plutarch's Isis and Osiris, 38.

12   There is no need to assume that the lunar calendar was used outside the priesthood, any more than the Catholic Easter calendar is used outside the Church.

13   Translation by R.O. Faulkner, 1993, Utterance 669, §1961.

14   In Greek mythology, the star of Cronus is the planet Saturn. On this, see Plutarch: The Face on the Moon, 941. Therefore, we may also see an association with Saturn in the Osiris Myth.

15   Plutarch, Isis and Osiris, 355; from the translation by F.C. Babbit.

16   We may see that: $360 \div 5 = 72$ rather than 70; but we should note that the same author, while discussing eclipses in his essay 'The Face on the Moon' says: "In fact the Egyptians, I think, say that the moon is one seventy-second part (of the earth)". If we take this as a misunderstood reference to the periods of the Moon's light then it suggests that Plutarch had intended to say one seventy-second part in his other essay also.

17   Hymn 1, 64: 11 & 15

18   We find something very similar in the Old Babylonian proclamation in the reign of Hammurabi, in which he declares "This year has a gap. Let the following month be called Elul II. Taxes must be paid to Babylon by the 25th of Elul II and not by the 25th of Teshrit." See Pomeras & Taton, pp 111-113.

19   The best conventional explanation for the sixty-year cycle seems to be that it was loosely based upon the Jupiter-Saturn conjunction, which repeats every 59.5779 years. On this see Needham, 1959, p408. The day

count would therefore be a sub-division of this.

20 See Needham, 1981, p 197.

21 One oracle bone refers to the renting of land from winter solstice to summer solstice of the following year; a period of 548 days. This suggests a measured year of 365.33 years.

22 Although the history was written much later, this refers to the period during or before the Shang dynasty. The Chinese at that period therefore believed that the year was 366 days long and were certainly aware that it was not 360 days. See Needham, 1981, p 102.

23 The mechanism of precession will be discussed in more detail in Chapters 9 and 10 below.

24 See Needham, 1959, vol III pp 245-246.

25 Translation by C. Waltham, p 5.

26 See Needham, 1959, pp200-201.

27 It may also be significant that 72 x 360 = 25,920 years; 100 x 260 = 26,000 years. This comes very close to the period of the precession of the equinoxes (see chapter 4 below).

4

The Great Cycle

No one has ever 'invented' a calendar. All calendars develop by the refinement of earlier ones. The Julian and Gregorian reforms, for example, required the once-off omission of several days or months yet were accepted with only a little protest. The population will adopt reform so long as the names and structure of the months and years remain recognisable to them. There may also be more complex rules, that are known only to an educated elite. One obvious example would be the paschal cycle used by the Catholic Church to calculate the date of Easter.

The adoption of a new *era* is quite a different matter. The simplest case arises when years are numbered from the reign of each new king, as in ancient Babylon and Egypt. A change of era involves no reform of the calendar, merely the numbering of the years, often associated with a new religion. The Christian era adopted by Rome in AD 532 was back calculated to begin at the supposed year of Christ's birth. Similarly, the Islamic world numbers its years from the flight of the prophet from Mecca in AD 622. For the present enquiry however, the most interesting eras are those that number the years from an ancient date, as we can therefore infer that the traditions preserve the earliest beliefs. We should beware however, of artificial eras and cycles. An obvious example here would be the Julian era adopted by modern astronomers.

Venus and the Mayan Era

When we look at ancient calendars and historical records, we can see that the priests and astrologers had an extraordinary ability to calculate the dates of important festivals by complex, and often arcane, arithmetic. The era of the Mayan calendar is a prime example. It was deciphered by archaeologists from the Long Count dates found on Maya monuments.

These dates can be correlated with the Julio-Gregorian calendar because the 260-day almanac is still used by some remote Maya communities. Specialists have correlated this with dates from the early colonial period and back to the codices of the Maya classical period. The Long Count dates all seem to have begun at a date of 13 baktuns, or 13.0.0.0.0 on the Maya vigesimal system. This corresponded to day 4 of the 20-day cycle and day 8 of the 13-day cycle, or 4 Ahau, 8 Cumka of the 260-day Maya tzolkin. If one assumes that there has been no break in the sacred almanac, this would correspond to a start date of *8 September 3114 BC* on the Julian calendar.

What was the significance of this remote era? It was a day of crucial importance in Maya mythology. They believed that time was a repeating cycle and that when a great cycle of 13 baktuns (13 x 144,000 = 1,872,000 days) had elapsed according to their calendar, then the world would be destroyed and the next cycle would commence. The era of the calendar therefore corresponded to the date of the most recent rebirth of the world. We may see that the next crisis falls due on 23 December 2012 and that there must have been an earlier one around 8239 BC, corresponding to baktun zero. Theorists may argue about whether the era was back calculated, or whether the calendar was really devised at such a remote date.

The Dresden codex holds a table of eclipses and appearances of Venus covering a period of 104 years. It details the divinations associated with these phenomena and reveals the true motives behind Maya astronomy. They took great interest in the synodic revolution of Venus, believing it to be most potent at the time of heliacal rising after inferior conjunction.[1] Various evils would occur at this time: Maya glyphs show the Venus-god hurling spears at the Earth and inscriptions associate it with drought, misery and war. It was essential to predict the rising of Venus in order that the priests could make preventive sacrifices.[2] We should therefore consider Maya astronomy rather as astrology and not attribute to them knowledge that cannot be obtained by naked eye observation. They would spare no effort to predict the movements of their gods with a mathematical precision.

Even people who take little interest in astronomy will have noticed Venus at its brightest, as a morning or evening star. These appearances repeat in a period that fluctuates between 577 and 592 days, giving an average synodic period of 583.92 days. However, the Maya rounded it to

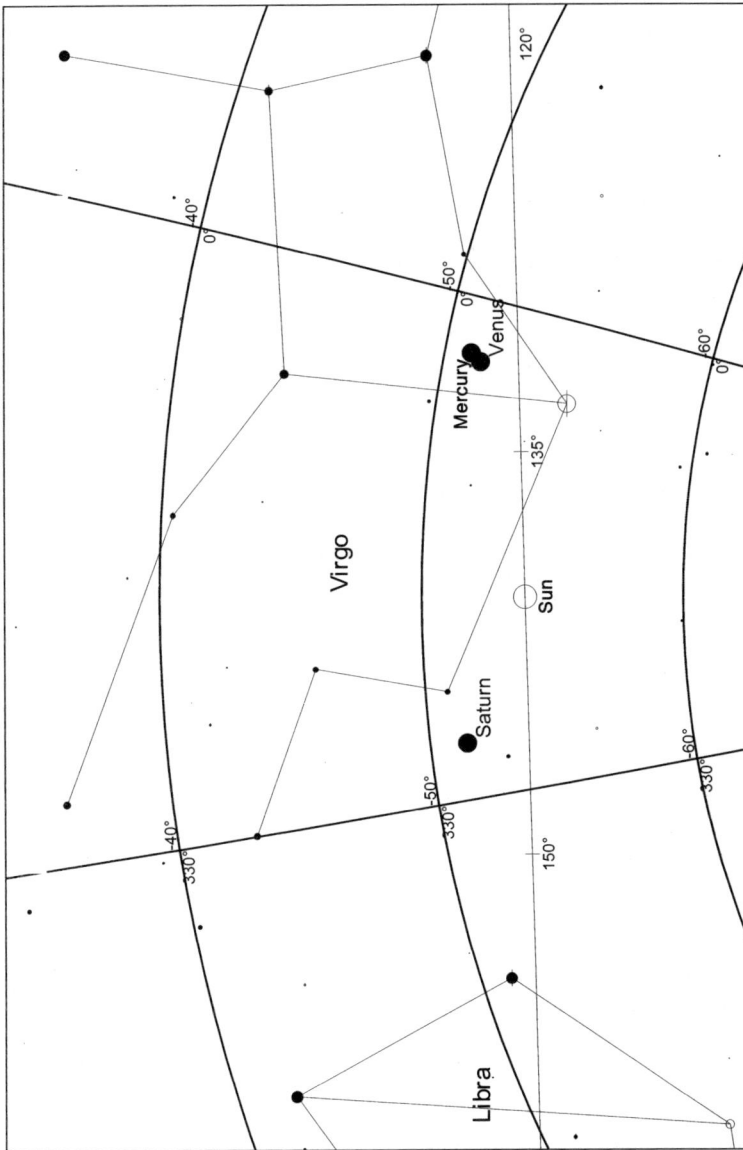

Figure 4.1 The Mayan Era
Planetary conjunctions at the era of the Mayan calendar (8 September 3114 BC). Venus was in close conjunction with Mercury and the Sun, with Saturn close by. Venus was the principle deity in Maya cosmology. (Retrocalculation using Skymap Pro 6.)

584 days. It happens that 5 x 584 = 8 x 365 = 2920. Using more precise values, 5 synodic periods last 2919.6 days while 8 solar years give 2921.9376 days; therefore Venus phenomena repeat 2.3376 days (2 days 8 hours) later each 8-year cycle. Moreover, every 24 years the cycle repeats almost exactly 7 days later in the year, with a difference of only 0.0128 of a day.

The Maya almanac in the Dresden codex charts Venus over 104 years. Now it may be seen that:

$$104 \times 365 = 146 \times 260 = 65 \times 584 = 37960 \text{ days}$$

The Maya were well aware that this relation was still just over five days longer than the real Venus, but they would never make any adjustment that would violate their 260-day sacred almanac. The Venus festival *must* always fall on the first day of the 20-day 'month' of Ahau, so the priests instead deducted 4 days at the end of every 61 cycles, to bring the real Venus back to 1 Ahau again. Even this was not considered accurate enough. The remaining error of just over a day was stored-up until it amounted to 4 days and then roughly every 481 years (676 x 260), they would deduct the 8 days. We may see that these rules are every bit as sophisticated as any rule of the modern Julio-Gregorian calendar – yet the Maya were a stone-age society.

Other parts of the Dresden Codex may contain comparable cycles for the planet Mars, but these remain to be fully understood. The eight-year cycle of 2920 days is also very close to the 2922 days (= 99 lunar months) of the Greek *octaeteris*; and some authors would therefore claim that the Venus table could be used to predict the occurrence of lunar eclipses.[3] Indeed, astronomer Anthony Aveni has suggested that the Venus cycles must originally have been calculated with respect to a lost lunar calendar, simply because the eight year concordance of Sun, Moon and Venus would be very obvious under such a calendar.[4] However, pursuing the rule that we will only investigate calendar cycles where they can cast some light on the meaning of mythology, these matters may be left for another time.

A Flood Catastrophe at 3100 BC?

We may see that the era of the Mayan calendar relates to a death and rebirth of the world consequent upon an astronomical event. They equated this with a particular aspect of Venus. We saw in Chapter 1 that a similar catastrophe may be recalled in the Celtic, Chinese, Indian and Babylonian mythologies, among others. Furthermore, the Babylonian Flood can be roughly dated to 3100 BC by mythological and archaeological arguments. The Mayan era of 3114 BC is therefore very intriguing. It may be that the event the Maya so feared was recorded as occurring at a rising of Venus, and that their era was back calculated based upon an orally transmitted legend.

Modern scholars have retro-calculated the start dates of Egyptian and Mesopotamian civilisations based on their king lists. Egypt and Babylon, as we have seen, did not use a calendar era, but numbered their years from the accession of each king. Their priests therefore had to reckon long periods by counting back these reigns in a manner not too dissimilar from modern archaeologists – except that the priests presumably had access to more contiguous records. This may tell us why every new Egyptian king was made to swear an oath, never to change the days of the year.[5]

Although it is often said that there is no extant Flood myth from Egypt (or indeed from anywhere in Africa) there are a few pseudo-Egyptian sources that suggest otherwise. These allow us to estimate a date for the Flood based on the Egyptian, rather than the Mesopotamian chronology.

The conventional chronology places Menes, the first king of Manetho's First Dynasty, at 3110 BC. Herodotus called him king Min, who apparently reigned at an era when the Nile flowed erratically.[6] He records that in the time of Min the present Nile Delta lay beneath the sea. Min is credited with the building a dam to control the Nile and founding the city of Memphis on the reclaimed land; and he was the first to unify Upper and Lower Egypt. Egyptologists equate him with a king Horus-Aha whose name is found in the oldest king list, the Palermo stone, dating from 2400 BC. Before Horus-Aha, the Palermo stone bears the names of kings who ruled in Upper and Lower Egypt before unification. Common sense dictates that we cannot consider a worldwide flood catastrophe more recently than the beginning of Egyptian history, which therefore forces us to seek a date before 3100 BC.

Diodorus Siculus thought that the southern Egyptians survived the Flood of Deucalion because of the aridity of their land.[7] Plutarch further tells us in his *Isis and Osiris* that all of Lower Egypt used to be an arm of the sea, until a deluge of rain forced out the sea and created the Delta. This event, he places at an era when Horus overpowered the god Set.

The Greeks equated Egyptian Set with their own god Typhon.[8] This god, as we have seen in Chapter 1, was a serpent-like creature who battled with the Titans in the age of Cronus and brought the Golden Age to an end.

The Myth of Osiris is preserved in its entirety only by Plutarch. He intimately associates the change in the length of the year, from 360 to 365 days, with the birth of the principal Egyptian gods. The five days were supposedly added to circumvent Re's curse, so that the five gods: Osiris, Horus, Isis, Set and Nephthys could be born. The equivalent Babylonian myth also has Marduk setting the length of the year following the sky-battle with Tiamat. Is it therefore possible to pin-down the date of this cataclysm by reference to purely Egyptian chronology?

In the surviving fragments of Manetho's *Aegiptiaca*, which are the basis for Egyptian history, we are told that the dynasties of mortal kings were preceded by dynasties of gods and demigods who reigned for a total of 24,925 years since creation.[9] Now here it must be cautioned that no full text of Manetho survives. It comes to us only in excerpts made by the Jewish historian Josephus and the Christian chronographers Syncellus and Eusebius, who were, of course, steeped in the Biblical chronology of creation. They dismissed the fabulous period of years, as a count of lunar months of thirty days. Eusebius for example, reduced the entire mythical period to 2242 solar years.

A similarly fantastic long chronology is preserved in the so-called *Ancient Chronicle*.[10] This shows demi-gods reigning for a term of 217 years before Menes, preceded by an era of the Gods lasting 984 years. Before this, the Sun ruled alone for a period of 33,000 years (= 55 Babylonian saroi). Together with the period of 2324 years assigned to all the mortal kings, this gives a total of 36525 years, corresponding to 25 Egyptian Sothic cycles (= 25 x 1461 years). We may therefore deduce that the chronicle was influenced by astrological theories; and that it was probably devised around 750-800 BC.

As Osiris, Isis and Horus were listed among the last of the gods to rule over Egypt, we may use this date to estimate the date when the Egyptians believed that Horus had been on earth. We may thereby arrive at a pseudo-historical date for the Flood. Given the modern range of uncertainty attached to the reign of Menes, we may place the era of Horus and Set between 3327 BC and 3167 BC.

Another way to get to this date is the chronology supplied by Plato for the destruction of Atlantis. Around 590 BC the Greek sage Solon visited Egypt and engaged in a discourse about the flood of Deucalion with two Egyptian priests. They argued that there had in fact been many such floods. Their own sacred records told them that the institutions of Egypt were eight thousand years old; and that a thousand years before that, the Atlantic island had been submerged in a flood catastrophe in the course of a single day and a night.[11]

Firstly, we have to take the Atlantis story at face value and accept it as just another Flood myth, but one of *Egyptian* rather than Greek origin. Although most commentators have taken the story to indicate the demise of a fabulous civilisation some ten thousand years ago, at the end of the Ice Age, it is in fact perfectly consistent with the other Egyptian sources. We may presume that the priest had before him a chronology much like that found in Manetho and the Old Chronicle, which gave a list of kings and priests, preceded by mythically long reigns of gods and demigods. Knowing that the catastrophe predated the Egyptian State, he therefore included part of the era of the demi-gods within his estimate. We may therefore more realistically place the flood-cataclysm during the predynastic era, at some time just before 3100 BC.[12]

Indian Cycles and Eras

Hindu tradition has preserved some of the oldest traditions in the world and these are intimately associated with the Indian calendar. The *Sûrya Siddhânta*, or 'solution of the sun', is believed to be the oldest and best of the astronomical solutions. Although the extant version was re-edited around AD 500, its content indicates that it may date from the sixth century BC.

According to the cosmology of the Sûrya Siddhânta, the stars revolve around a cosmic mountain called *Meru* that supports the sky; a concept not unlike that found in Chinese cosmology. The gods dwelt on its summit, much as the Greek gods dwelt on the summit of Olympus. The passage of a year for human beings was merely a divine day for the gods. Spring and summer comprised the divine day, in which the sun shone more brightly upon the gods, while autumn and winter were the divine night.

The divine cycle of time was based upon the important Brahmanic numbers 10,800 and 432,000. These numbers occur in the epic Rig-Veda, which is organised into 10,800 divisions of forty syllables each, giving a total of 432,000 syllables. These same divisions were embodied in the calendar, thus giving us a clue that the system is at least as old as the Vedic period (c 1500 BC). The Rig-Veda contains as many metres as there are moments in the divine year:

1 year	=	12 months
1 month	=	30 nychthemera
1 nychthemeron	=	30 moments

A human day or nychthemeron consists of fifteen moments of daylight and fifteen of darkness. It may be seen that 12 x 30 x 30 = 10,800 and that each divine moment corresponds to 48 minutes of time.

Thus, each human year was but a day for the gods and 360 divine days comprised the divine year. These were further grouped into a Great Year of 12,000 divine years after which interval it was believed that the planets completed their revolutions and returned to their starting positions. It may be seen that:

12,000 x 360 = 10 x 432,000 = 4,320,000 solar years

Cycles of time were based upon the *yuga*, for which there are a variety of definitions.[13] The older Brahmanic tradition calls this Great Year the *mahayuga* and divides it into four successive yugas in the ratio 4:3:2:1 in which the world has descended into ever-greater disorder. Thus, the mahayuga was divided as follows:

1 x 432,000	=	432,000 years	=	kaliyuga
2 x 432,000	=	864,000 years	=	dvapurayuga
3 x 432,000	=	1,296,000 years	=	tritiyuga
4 x 432,000	=	1,728,000 years	=	krtiyuga
Total		=	4,320,000 years	= mahayuga

An even larger unit called *kalpa*, or a day of Brahma, consists of 1000 mahayuga, or 4.32 billion years – which comes remarkably close to modern estimates for the age of the Earth.[14]

In the Hindu view of time, Krti-age corresponds to the golden age of mankind, and the present age is the Kali-age in which human life is at its shortest. During the transition between each age, the universe is destroyed and lies submerged in the cosmic ocean while Brahma sleeps. With each new age, the world is created anew. Each creation corresponds to an avatar of the god Vishnu.[15]

This picture is further complicated by an alternative historical tradition found in the Mahabharata and the Puranas. The Puranic historians similarly attempted to divide the historical period into four yugas. This alternative chronology, based on generations of kings, would confuse the thinking of the later historians when they attempted to define the era of the Kaliyuga.

What has all this mythology to do with the calendar era? In the year 499 the Indian astronomer Aryabhata, using older astronomical sources, stated that he had lived for twenty-three years at the end of year 3600 of the kaliyuga. He redefined the mahayuga as four equal periods of 1,080,000 years, because astronomically (based on a slightly inaccurate calculation of the length of the year) it was the smallest number of days that comprised an integral number of solar years. He defined the era of the Kaliyuga as 3600 years before AD 499, which corresponds to midnight at Lanka on *Friday 18 February 3102 BC*.[16]

Taken together with the mythology, this astronomically calculated date gives us yet another way to date the Flood. The era of the Kaliyuga corresponds to a date when the world was supposedly destroyed by fire and water and recreated anew. Viewed in a purely Indian context the date is not remarkable, but the correspondence that we may find with other world flood myths is striking.

Unfortunately, we cannot know how this date came to be significant. It is generally assumed that the Siddhantic astronomers back calculated this era based on the observed periods of the planets. It was believed that the era commenced when all the planets were aligned on the celestial equator in the zodiacal sign of Mesha (Aries). In other words, when all the planets were hidden behind a total eclipse of the Sun occurring exactly at the spring equinox. Unfortunately, such theoretical back calculation can never be exact because the planets perturb each other. However, a modern retrocalculation of the planetary positions at the era of the kaliyuga does indeed show a close arrangement of the Moon and planets with the Sun, in the constellation Aries (see figure 4.2).

As with the Mayan calendar, it may be that oral tradition preserved a memory that a world-wide catastrophe had occurred at a time when all the planets were in close conjunction with the Sun; and that the date of this conjunction was derived subsequently by astrologers in both India and Babylon.

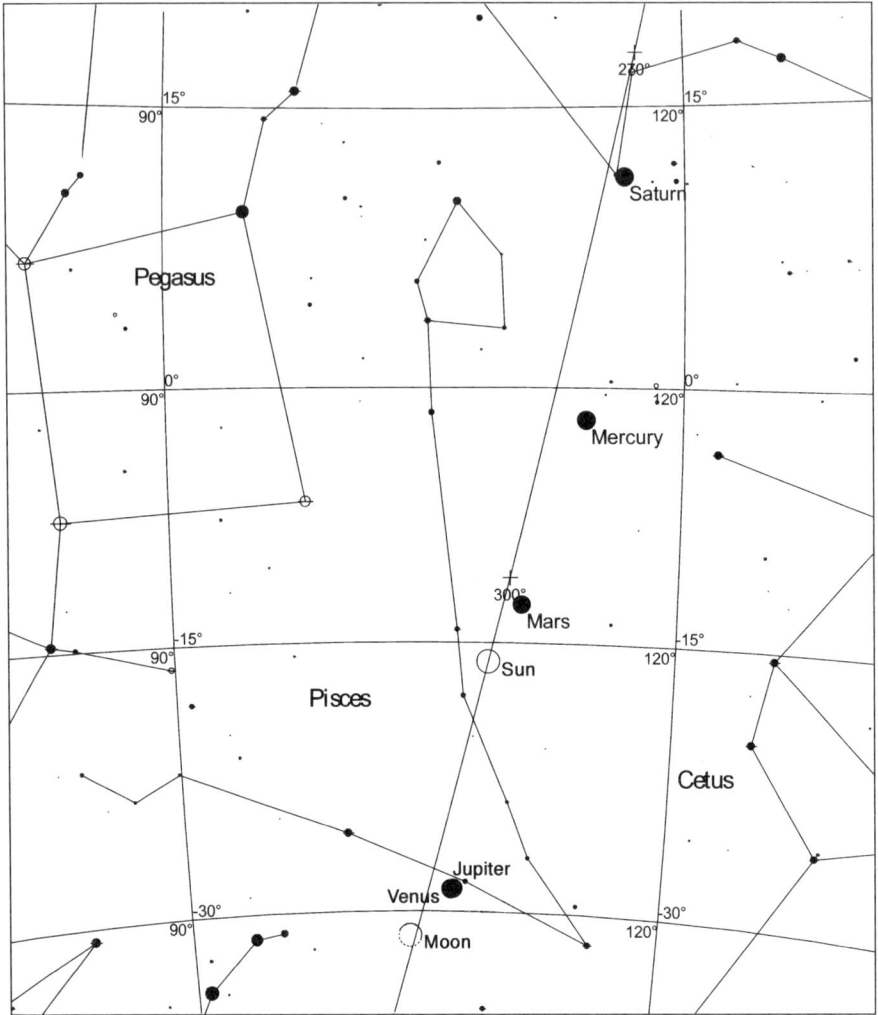

Figure 4.2 The Kaliyuga
Planetary conjunctions at the era of the Kaliyuga (February 18 3102 BC). A close conjunction of Venus Jupiter and the Sun occurred in the constellations Aries and Pisces, with the Moon and the three other bright planets close by. Indian cosmology taught that all the planets were in perfect conjunction with the Sun at the start of Kaliyuga.
(Retrocalculation using Skymap Pro 6).

Traditions of the Great Year

Multiples of the Brahmanic numbers 108 and 432 occur surprisingly often in other astronomical contexts. As we have seen, the period of the kaliyuga, or 432,000 years, is also encountered in the Babylonian chronology of Berossus, where it represents a period of 120 saroi, or 1200 years of 360 days. It is unfortunate that we do not possess the full text of Berossus, but it would seem reasonable to suggest that at some point his chronology of kings and demigods was expanded to fill out a theoretical period of creation derived from purely astrological theories.

We may perhaps see why some of the long chronologies given for kings before the Flood have arisen. Because these kings were viewed as demigods, their reigns were to be measured in *divine*, rather than human years. Josephus accounted for the long lifetime of Noah in much this way, saying:

> ...God permitted them [*the ancients*] to live longer because of their excellent character and the usefulness of their discoveries, since unless they lived six hundred years — for so long is the period of a great year — they could not have made accurate predictions... For Manetho, the historian of Egypt, Berossus, the compiler of the Chaldaika...all agree...In addition Hesiod, Hecataeus...and Nicolaus record that the ancients lived a thousand years.[17]

From the text of the Indian Sûrya Siddhânta we are also told that the period of a Mahayuga corresponded to 146,568 revolutions of the planet Saturn.

$$4,320,000 \div 146,568 = 29.4744 \text{ years}$$

The true sidereal period of Saturn is 29.456 years, or 10,760 days. So may we see in this, an attempt to define a Great Year by the period in which Saturn and the Sun return to their starting positions? It must be remembered that all ancient astronomy was geocentric. The period of Saturn's revolution as viewed from Earth is 378.1 days giving roughly 10,965 days in 29 synodic periods. Thirty solar years give 10,957.5 days, whereas 30 x 360 =10,800 days.

These are again extraordinary coincidences and it would seem that the number 10,800 must also be associated with some ancient attempt to determine the orbit of Saturn based on a 360-day year. A human year is but a divine day of the gods and so if we multiply the thirty-year orbit of Saturn by 360 we arrive at the period of the divine year.

30 x 360 = 10,800 human years

This same philosophy seems to have found its way west to ancient Greece, where we find a similar theory among the fragments of the philosopher Heraclitus (c500 BC). A passage of Aetius reveals:

> Heraclitus said that [the Great Year consists of] 10,800 solar years. This year is also called 'solar' by some and by others 'God's Year'...Heraclitus and Linus thought it occurred after the passage of 10,800 years.[18]

Other fragments of this same philosopher reveal his belief that in accordance with particular cycles of time the universe is consumed by fire and reborn;[19] and that the period of a human generation should be

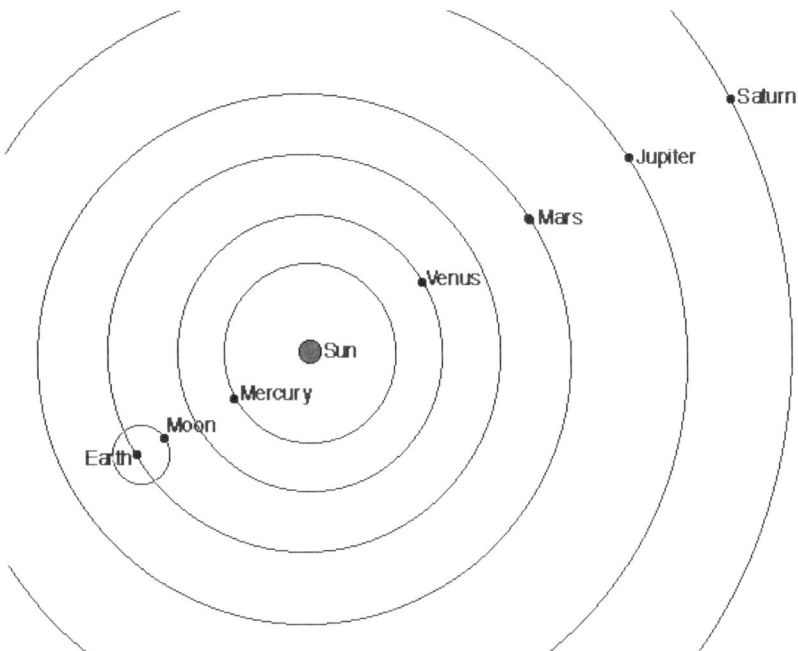

Figure 4.3 The Grand Conjunction of Planets.
For ancient observers it would not matter whether Venus and Mercury were at superior or inferior conjunction.

reckoned as thirty years.[20] Thus, in keeping track of Saturn's orbit we also count the human lifetime.

According to Diodorus, the Babylonian astrologers referred to the five planets as 'interpreters' in that they foretold to mankind the will of the gods, but they were particularly mindful of the movements of Saturn.

> ...but if referring to them singly, the one named Cronus by the Greeks, which is the most conspicuous and presages more events and such as are of greater importance than the others...[21]

Berossus supplies the Babylonian definition of the Great Year. He was so confident of his astrology that he was even prepared to predict a definite date for the final catastrophe, believing that this was influenced by the course of the planets. He asserted that when all the planets assembled in the constellation Cancer (midsummer), such that a straight line would pass through their positions, then the world would be consumed by fire. When the same arrangement occurred in Capricorn (midwinter) then it signified the deluge.[22] This is surely giving us a version of Indian cosmology, but whether it truly originated there, or in Babylon, Egypt or elsewhere, we may never know.

Chinese Resonances

Chinese philosophy recognised the existence of a celestial sphere, as embodied in the *Kai Thien* cosmology (literally: 'heavenly cover'). It perceived the sky as a hemispherical bowl that floated on the ocean and rotated about a great mountain. The Earth by contrast was viewed as a square bowl, with its corners set at the four cardinal points. In the very earliest tradition, these were the four feet of a tortoise, which carried the world on its back. The Canon of Yao (before 2255 BC) describes how the mythical Brothers Hsi and Brothers Ho were despatched to these four extremities to regulate the occurrence of the seasons. Astronomers of the later Han Dynasty would further extend the bowl into a true sphere and recognise the existence of a south pole as well as a north pole.

The origins of the Kai Thien cosmology are undeniably ancient, for we find these same notions embodied in the Chinese Flood myths. It arose as a by-product of the measurement of the solstice shadow. According to tradition, its originator was the legendary hero Fu-Hsi, who devised the system of degrees to divide the sky for calendrical purposes.[23]

The pillars of the sky were situated atop the mountain *Pu-chou Shan*, the 'imperfect mountain', so-called since Kung Kung struck one of the pillars and caused the sky to lean over.[24] This mythology intimately links the tilting of the axis of rotation with the Flood, as facets of the same event. A legendary pseudo-chronology places Kung Kung in the reign of an empress called Nu-huang, or Nü-Wa, who supposedly ruled between 2953BC and 2838BC. Chinese historians have always treated these dates as unreliable, but there are few other ways to estimate the date of the Flood based on purely Chinese sources.

Although Chinese astrology knew a world-cycle similar to the Indian model, it seems they never adopted it as an era. Astronomers of the Han Dynasty believed that the current world-cycle had commenced with a grand conjunction of the planets and that it would, one-day, end in the same way. The antiquity of this belief may be seen in the sexagesimal cycle.

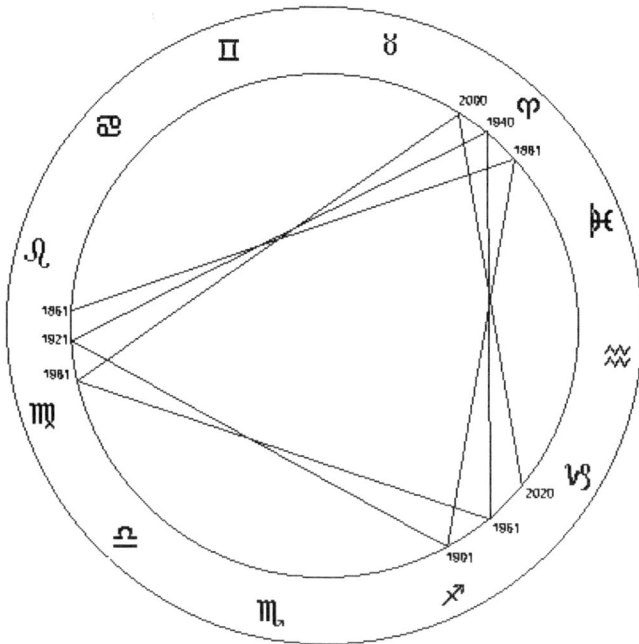

Figure 4.4 *The Trigon of Jupiter and Saturn.*
Roughly every sixty years the Jupiter-Saturn conjunction returns to the same area of the sky.

One theory is that the twelve-year animal cycle was originally intended to track the twelve-year orbit of Jupiter (more precisely 11.86 years) and that the sixty-year cycle tracks the Jupiter-Saturn conjunction. This rhythm arises because two Saturn revolutions of 60 years very nearly correspond to five Jupiter revolutions of 12 years. Jupiter-Saturn conjunctions occur roughly every 20 years. Moreover, every third conjunction occurs at a similar position in the zodiac, displaced only 8° to the east. This theory is fine as far as it goes, but as the more precise period of the conjunction is 19.86 years, such a calendar, like any other, would require periodic intercalation to maintain its accuracy. There is no evidence that this was ever done.

The later Han astronomers were well aware of the luni-solar correspondences such as the 19-year *chang* and the 76-year *pu*; and the eclipse cycle was closely watched from the earliest times. They combined these cycles with the planetary periods to arrive at the period of Grand Conjunction and calculated that all the planets would repeat their motions in a period of 138,240 years. This number is a multiple of 8, 12, 30, 60 and 360 (though not of 365).

The Chinese have been predicting and observing eclipses as astrological portents since at least 1500 BC, as is proved by the Shang oracle bones; and a passage in the Shu Ching may even carry this astronomy back as far as the third millennium BC. Their eclipse cycle was based on a period of 135 lunations, which they multiplied by 47 to give a period called a *hui*. This is also quite close to the period of the Mars-Jupiter-Saturn conjunction, which recurs in the same part of the sky every 516.33 years.[25]

A still larger multiple was made by a multiple of twenty pu, or 1520 years, which was also a multiple of the sexagesimal day count.[26] The multiples are as follows:

20 x 76 = 1520 solar years

This was further meshed with the sixty-year cycle to make a great cycle called the *chi*.

21 x 1520 = 31,920 years
31,920 x 365.25 = 11,658,780 days
21 x 19 x 487 x 60 = 11,658,780 days

The Han astronomers believed that this cycle of 31,920 years was the period in which world events repeated and so its role is comparable to the long cycles encountered in Babylonian and Egyptian chronologies. They also attempted to calculate the grand conjunction of the planets based upon multiples of the *measured* length of the year and the lunation.[27]

The most important lesson from Chinese astronomy is that they certainly knew the lunisolar correspondences, upon which western calendars were based, yet they apparently did not employ them. They were just as obsessive in predicting and observing eclipses as were the Babylonians. Yet, they preferred to base their calendar upon contemporary measurement rather than upon ancient traditions. Chinese astronomy is therefore much less useful as a tool for probing the ancient past.

The Significance of the 108 and 432 day cycles

We may easily recognise a 360-day cycle as a vague approximation to the solar year and the number 30 as a standardised month, but multiples of 432 and 108 are less obvious. What could these numbers imply?

The first astronomical phenomenon to eliminate from the discussion is the *precession of the equinoxes*. Due to a gravitational couple, the Earth's inclined axis is forced to make a conical motion about the pole of the ecliptic with a period of 25,920 years. This equates to 1° every 72 years and so is barely noticeable over a human lifetime. Yet despite its long period, it is significant enough to have been noticed by any ancient culture that maintained star charts. It is therefore possible that some of the long-cycles found in astrological texts, such as the divine era of 24,925 years found in Manetho, may be based on an ancient estimate of the precession.

In the West, the discovery of precession is attributed to Hipparchus around 126 BC, who calculated that the stars had shifted by about 2° to the east relative to the equinoctial points, based on the older star charts of Timocharis.[28] At the time of its discovery, the spring equinox occurred in the constellation Aries. Astronomers still call this the 'First Point of Aries' and treat it as the origin for stellar co-ordinates – although it has now precessed into the neighbouring constellation of Pisces. Between 4400 BC and 2200 BC the vernal equinox would have occurred in Taurus and the autumn equinox in Scorpio.

The Han Chinese astronomer Yu Hsi independently discovered the precession in the Fourth Century AD and it presumably played no part in earlier Chinese thinking.[29] However, in India the problem was well known, if not entirely understood. The oldest calendrical references in the Vedas listed the first nakshatra as the Pleiades star cluster in Taurus; a later list began at Aries, a still more recent list began with Taurus again, while the Siddhantic astronomers observed the equinox to occur in Aries. Now we may deduce that one view must have been based upon very old traditions (dating from the third millennium BC or earlier) while the other was based on contemporary observation.

This completely confused the Indian astronomers, who concluded that the equinox 'librated' or oscillated between the two constellations. They adopted the rate of the motion as 54 seconds per year, which is half of the magical number 108.[30] The true average annual rate is nearer 50 seconds. The earliest Indian astronomers must surely have noticed the precession because of the way their calendar operates, but whether they tried to embody it in the number 108 and its multiples is open to debate. It is certain though, that this reliance on tradition fossilised Indian astronomy in a way that could never occur in China.

What then could be the significance of the number 432? We are given a clue by the Chinese myth of Kung Kung, which suggests that the Earth's axis tilted over at the time of the Deluge. If, for the sake of argument again, we accept the myth at face value then we must be clear that this would represent a real change to the angular momentum of the Earth. Any such shift could only be caused by a strong external force, perhaps a comet impact. Unless the force were applied precisely at one of the poles, or exactly along the equator, an impact must divide its energy partly in changing the obliquity, and partly in changing the length of day. A change to the length of day should reveal itself in our oldest calendar traditions, and as we have seen, these indeed suggest a former year with 360 days.

Now, modern geophysical theory tells us that if the diurnal rotation is disturbed then it must wobble in two natural modes before it can settle to a new state. These will be discussed in more detail in later chapters. For the present it will suffice to summarise that one mode has a period of about 435 days and is short-lived, being damped after about 20 years; the second has a more uncertain period estimated at 440 days and should have a much longer lifetime, perhaps as long as 1500–2500 years – but these should only be considered approximations.[31] It may be seen therefore that if the myths of the Great Flood recall a real impact event in the years before 3100 BC then the rotation would still have been wobbling as recently as 1500-500 BC. This coincides with the period from which we find evidence of the earliest calendars. Therefore, could the number 432 be an attempt, by the astronomers in ancient Babylon and India, to schedule this phenomenon in the calendar?

In 1960 the Australian astronomer George F Dodwell wrote to the author Rene Norbergen, who had just returned from an expedition to Turkey in search of the remains of Noah's Ark. Norbergen included Dodwell's comments in his book *Secrets of the Lost Races*.[32] As one of the most eminent Southern Hemisphere astronomers of the twentieth century, Dodwell, had spent some twenty-six years investigating the secular variation of the ecliptic. Based on observations of the Sun at the solstices over the past 3,000 years he concluded:

> ...I find a curve, which after allowing for all known changes, shows a typical exponential curve of recovery of the earth's axis after a sudden change from a former nearly vertical position to an inclination of 26°, during the interval of the succeeding 3,194 years to AD 1850...The date of the change in the earth's axis, 2345 BC.

Dodwell concluded that this event was none other than the Biblical Flood and that the story of Noah was true.[33]

An exponential recovery curve is indeed what would be expected if the axis of rotation were displaced from its position of rest. During the early stages the motion would be very obvious as a spiralling of the pole of the heavens as it hunted about the rest position — the new celestial pole. The motion would betray itself in the length of the solstice shadow and in the rising and setting positions of the Sun and Moon; and would surely have been noticed by the ancient astrologers. If the motion were significant, it would also exhibit seasonal climate effects. A 440-day period would combine with the year to give a long-seasonal rhythm of roughly seven years.[34]

It may be that a period of 432 days was chosen because the number also had other useful properties, as we have seen above. Ancient thinking was entirely astrological and unscientific. Perhaps they tried to explain the seasonal variations as the influence of the planets and deduced a heavenly control over human affairs that is entirely coincidence. In any event, as the motion was transient the original meaning of the number 432 would be forgotten and became shrouded in religious dogma.[35]

Do we possess any other evidence that such a spiralling motion of the celestial pole was observed in ancient times? Perhaps surprisingly, the answer is *yes*, but it has always been interpreted as a precocious knowledge of the precession of the equinoxes.

The Chinese *Kai Thien* cosmology was allied to a theory that explained the seasons by a rising and sinking of the world on the bearings of the polar axis. The solstices were explained because as the pole rose on its bearings in summer it grew closer to the Sun. It winter it sank and was further from the Sun. However, there is no mention of any irregularity in this mechanism.

Some commentators have compared the Kai Thien cosmology to similar concepts in the Timaeus of Plato, which reflects the earliest Greek cosmology before it was overridden by that of Ptolemy and Aristotle. In a notoriously obscure passage, Plato describes the Earth as 'winding about the axis of the universe'.[36] As he has already accounted for the daily rotation of the fixed stars and the motions of the planets, this seems to imply an additional motion of the axis. Many scholars have debated what Plato intended and have translated this word variously as 'oscillation' or 'rolling'.[37] While it is difficult to see it as a sliding motion (as in the Kai Thien model) it would be, however, a fair description of the Earth wobbling slightly on its axis.[38]

A few centuries after Plato, Aristotle would assert that the Earth did not change its position in any way whatsoever and western astronomy took a different path. It is easy to see therefore, why the Neoplatonists would later term the precession of the equinoxes as 'the Platonic Year' although the precession was not discovered until some two centuries after Plato's time.

Understanding the Ancient Sky

In our earliest evidence we therefore find a contradictory view of the competence of ancient astronomers. We see that astronomers in all of the world's earliest cultures were capable of devising the most complex cycles and astrological theories. We see long periods of years being counted by a complex meshing of planetary cycles with the year and the month – yet scholars would have us believe that these same astronomers could not perform the relatively simple measurement of the length of the year.

Whenever we can look back to the second and third millennia BC, we see the preference for lunar rather than solar calendars. We see that, certainly in China, they knew the lunisolar correspondences on which modern calendars are based – yet they preferred to measure the year and the month. Similarly, in Babylon and India, we see that they were directly observing the start of each month rather than relying on simple multiples of day-counts. In Egypt, we see that the priests measured the true length of the year, yet they obstinately preferred to let their solar calendar wander through the seasons.

We should ask *why* all the ancient civilisations of the world seem to have believed that the period of the year was 360 days. It will not do to dismiss this as an inaccurate or approximate period for the year. Archaeologists are adamant that the civilisations of the Americas evolved completely independently of the Old World, yet they too evolved a year of 360 days.

We should also ask why all the earliest cultures of the world seem to have believed that the world had been destroyed, by fire, water or both, at a date close to 3100 BC? Scholars deny that this convergence (if indeed they recognise it at all) could be due to ancient contact between civilisations, or indeed a legacy of wisdom from some archaic 'lost civilisation'. Perhaps the convergence of traditions is best explained by the common experience of a real cataclysm; and the independent development of calendars and chronologies by generations of astrologer-priests, living under the same sky and observing the same phenomena. Great minds, we are told, think alike, and this is surely true of the earliest astronomers.

References to Chapter Four

1   The synodic revolution of a planet is the period of the geocentric orbit, usually given as the period between two successive conjunctions of the planet with the Sun as viewed from the Earth.

2   Thompson, J.E.S. (1974) p 87

3   Aveni, A. (1997) pp122-5

4   Five solar years = 2921.93d; ninety-nine lunar months = 2923.47d; five Venus cycles = 2919.6d

5   Parker, R.A.(1950) p 54

6   Herodotus, II, 99-100

7   Diodorus Siculus, 1, 10, 2-11

8   Plutarch, Isis and Osiris, 367

9   Manetho Aegyptiaca (Epitome) Fragment 1

10 See W.G. Waddel, 1940, Appendices 3 & 4. One version is preserved via Manetho's Book of Sothis (or 'The Sothic Cycle') and another via the chronographer Panodorus.

11 In other words, in the course of a single rotation.

12 In *The Atlantis Researches*, I arrived at a similar date using the statements of Herodotus (Histories II, 42) which would place the foundation of Egypt

at two complete Sothic cycles prior to his visit to Egypt.

13 Later tradition would give the yuga as a period of 1830 days [(5 x 360) + 30].

14 Another large unit is the Manvantera or year of Manu, defined as a period of 71 mahayuga or 306,720,000 years.

15 It seems likely that the myth of the ages is far older than the worship of Vishnu and was originally associated with Prajapati in the Brahmanic tradition. See O'Flaherty, 1975, pp 179-185.

16 Lanka is the theoretical position in the mid-Indian Ocean where the meridian passing through Ujjain in India crosses the equator. See Koshar, R, 2000 pp 46-48.

17 Josephus, Jewish Antiquities1, 107 from the translation by T.M. Robinson, p 29

18 Aetius 2.32.3; translation by T.M. Robinson, 1978, p 176

19 See Robinson, p174

20 Plutarch, Obsolescence of Oracles, 415

21 Diodorus Siculus, II, 30; from the translation by C.H. Oldfather

22 See Robinson, p 15

23 Needham, 1959, p213

24 See above, Chapter 1

25 The multiples are as follows:
47 x 135 x 29.53
= 187367.85 days
= 512.99 solar years
513 x 365.2422
= 187369.2486 days
20 x 76 x 365.25
= 555,180 days
19 x 487 x 60
= 555,180 days

27 The Han astronomers arrived at a period of 138,240 years in which all the planetary motions repeated; and a period of 23,639,040 years called the 'Supreme Ultimate Grand Origin'. Fascinating though these figures are, there is no reason to believe that they preserve any ancient knowledge.

28 Ptolemy, Syntaxis, VII, 2

29 Needham, 1959, p 200

30 Pomeras & Taton, 1963, pp 144-145

31 See below, Chapter 10

32 Norbergen, 1977, pp 20-21

33 Dodwell's unpublished manuscript "The Truth of the Bible" is in the Barr Smith Library at the University of Adelaide. I have been unable to investigate his research further, as the family would not allow the papers to be copied and I was not prepared to travel to Australia to view them.

34 I have investigated this seven-year correspondence in more detail in *The Atlantis Researches*. Seven-year climatic variations are mentioned in the Biblical story of Joseph, in the Epic of Gilgamesh and in some Celtic myths.

35 Also note that under the regime of a 360-day year the period becomes 440 x 360/365.25 = 433.67 days, which allowing for the margin of error in modern estimates, is close enough to 432.

36 Plato, Timaeus, 40b

37 Heath, Sir Thomas, 1932, introduction, xli.

38 However, in Epinomis, 990 (see Introduction) Plato makes no mention of this extra motion.

5

A Five and Six of Years

Ancient Europe offers us few opportunities to examine the earliest calendrical science. As we saw in Chapter Three, the calendars of Classical Greece and Rome were unreliable lunar calendars that were superseded by imported wisdom from Babylon and Egypt as soon as it became available. We cannot know what the calendar practice was before the early classical period and certainly nothing that would take us back as far as 3000 BC.

Elsewhere in Europe, we have little more than oral tradition to go on. However, we do have one clue to a calendar that is certainly 'antediluvian' and that is found in Plato's myth of Atlantis.[1] He tells us that, at some remote period, before the geological catastrophe that we may equate with the Flood, the kings of the various parts of Europe gathered for a religious ceremony:

> ...every fifth and every sixth year alternately, thus giving equal honour to the odd and to the even number.

Now it is this passage more than any other that convinces the present author of the authenticity of the Atlantis myth.[2] If we take it at its face value, it is a piece of mythology transmitted to us from Egypt, via the Greek sage Solon. Where have these alternated five and six-year periods come from? The Egyptian calendar as we have seen was based on observation; the Babylonian calendar upon the nineteen-year intercalation cycle; and the Greek calendar of Solon's day, upon an eight-year cycle. A cycle of five plus six years implies a calendar with an *eleven-year* intercalation cycle and four intercalary months (see table 3.1).

Remarkably, it is indeed possible to reconstruct a highly accurate calendar based on these principles. If the Atlantis myth were merely

Plato's fiction, as most scholars seem to believe, then why would he go to so much trouble to invent a completely new calendar based upon an eleven-year cycle – a cycle which, so far as is known, has never been used anywhere else in the world? Moreover, why does Plato not mention the five-plus-six-year calendar in any of his astronomical works?

There is a very similar calendar formula in one of the songs of the Finnish epic *Kalevala*.[3] Here we find the maid Marjatta lamenting the passage of a long period of years as, "for five or six of summers". The use of this five-plus-six year formula in Finno-Ugrian oral tradition gives us a strong clue that this cycle has a very ancient origin in prehistoric Europe.

The Celtic Calendar of Coligny

In Julius Caesar's account of his Gallic wars, between the years 58 and 51 BC, we find a brief account of the activities of those priestly philosophers of the Celts known as Druids. Very little is really known about their science; Caesar apart, we have only a few passing references in the works of other classical writers. Amongst the scant information that survives is the recognition that they were highly capable astronomers, having preserved this knowledge orally through many generations. Caesar tells us that the Gaulish Druids held long debates about the size of the Earth and the movements of the planets.[4]

The Romans considered the Druids a threat to their authority and ruthlessly suppressed their order in Gaul; particularly during the reigns of Tiberius and Claudius. We must presume that they also suppressed it within their British province. Indeed the Druids' continuing interference in Gaul may have been one of the underlying motives behind the Claudian conquest of AD 43. It is Caesar again, who tells us that the order originated in Britain; and that some kind of Druids College existed there.[5]

The most interesting of all the artefacts surviving from this period of suppression is the Coligny calendar; unearthed in 1897 in a vineyard near Bourg-en-Bresse. This remarkable document consists of several fragments, constituting some three-fifths of a bronze tablet, which appears to have been deliberately smashed. The reconstituted tablet forms part of a calendar, annotated in the Gaulish language, using the Roman alphabet; and this has been used to date it approximately to the reign of Augustus. The mechanism of the calendar shows no trace of Roman influence and it must therefore belong to the final period of native Gallic culture. A less well preserved fragment found near Villards d'Héria in France is thought to date from the second century AD, but it adds little to our knowledge of the calendar's workings.[6]

Although most discussion of the Coligny Calendar has treated it as a purely Gaulish artefact no older than the Roman era, it is important to state yet again that all calendar traditions, wherever we encounter them, have evolved from ancient roots. The first-century Gauls no more invented the Coligny Calendar than you have invented the calendar that hangs on your own wall! It may therefore hold evidence of astronomy that was performed in western Europe *thousands* of years before the extant calendar itself. Caesar's comment that Druidism was 'found existing in Britain'

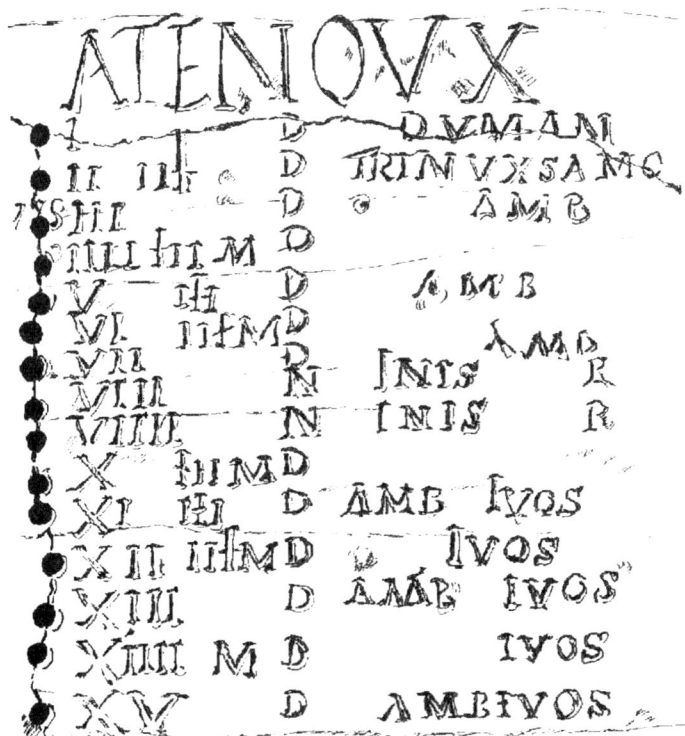

Figure 5.1 The Coligny Calendar
A drawing of part of the Coligny Calendar, showing the notation for the dark-half of the month Samos (=June). The notation TRINVXSAMO is probably the midsummer festival; the monthly festival IVOS shown here, exhibits intercalary displacement in years 1 and 4 of the calendar.

would suggest that we should look to the British Isles for the origin of the Celtic calendar traditions.

There can be little doubt that the Coligny Calendar was a lunar calendar, intended for public display. The tablet holds notation for five years; set out in sixteen columns. Each column holds the notation for four named months; the exception being the first and the ninth columns, which hold an intercalary month, followed by two normal months. The intercalations are therefore positioned at the beginning of the five year cycle and in the middle of the third year (i.e. year one and year three were increased to thirteen months). Hence, it may be seen that the intercalations were spaced at intervals of two-and-a-half years.

With a single exception, the length of each month is always either 29 or 30 days. Each month is marked-out in days, listing their various festivals; and with a peg-hole for each day so that the date could be marked. Since the average period of the lunar month is 29.53 days, any lunar calendar must incorporate alternating 29 and 30-day months. Furthermore, since even this allows a divergence to develop, one of the months must also be variable (equivalent to our February) and used to regulate the correspondence with the real Moon. In the Coligny Calendar, the variable month is called Equos ('horse-month'). The 30-day months are each suffixed as 'good', whereas the 29-day months are styled 'not-good'. As Equos was similarly 'not-good', this indicates that it was not considered an ordinary thirty-day month. It may have alternated in length between 28 and 30 days.

As the extant calendar is incomplete there is insufficient information to tell us conclusively how it operated. The surviving fragment contains notation for only sixty-two months and covers a period of approximately five solar years. Sixty-two lunar months contain 1831 days; whereas five solar years require only 1826 days. The sixty-two months exhibit notation for a possible 1835 days. Therefore, if the Coligny fragment were to constitute the entire calendar then, by lunar reckoning, it would suffer a cumulative error of four days in every five years; and an error of nine days by solar reckoning. Such a huge error defeats the purpose of the intercalary months. There must therefore have been a correction mechanism or another, missing, cycle alternated with this one, in order to pull back the divergence.

If we examine the notation of the Coligny calendar in more detail, it is evident that each month was also divided into two halves, or 'fortnights'. The second half of each month is headed ATENOUX, which means 'returning night'.[7] We may therefore presume that the full moon was intended to fall in the middle of the first fortnight. However, it is impossible to make the full moon fall on a precise day in every month and some allowance must be made for the appearances to fall a day or so either side of the ideal calendar day.

For a calendar to be useful to the general population, it must be organised according to simple rules that everyone will remember. The best way to do this would be to use one of the self-correcting lunisolar cycles. A theory that the Celtic calendar was based upon the nineteen-year Meton cycle was proposed by some of the first scholars to investigate the Coligny calendar.[8] However, there is no historical evidence that such a cycle was ever used by the ancient Celts, either in Gaul or in the British Isles. Later commentators have therefore preferred to reconstruct the Druids' calendar, based on a cycle of thirty years, for which, as we shall see, there is more solid documentary evidence.[9]

The Hyperboreans

We do find a reference to the Metonic cycle in the passage of Diodorus Siculus concerning the northern *Hyperboreans,* which we met in Chapter One, and which he attributed to Hecataeus of Abdera (fourth century BC). Hecataeus locates these people in an island opposite the coast of Gaul, which must therefore imply either Britain or Ireland.

> In this island there is a magnificent sacred grove of Apollo and a remarkable temple of a round form...that the god visits the island once in a course of *nineteen years* in which period *the stars complete their revolutions* [present author's italics] and that for this reason, the Greeks distinguish the cycle of nineteen years, by the name of the Great Year. During the season of his appearance, the god plays upon the harp and dances every night from the Vernal equinox, until the rising of the Pleiades...[10]

This passage will be further discussed below in its proper place. For the moment, it should be noted that it has been influential to many theories of astronomical alignments at Stonehenge and other stone circles throughout the British Isles. This goes back to the work of Hawkins and others in the 1960's, which triggered an explosion of interest in this field.[11] A leap of faith is required to accept that the Hyperboreans

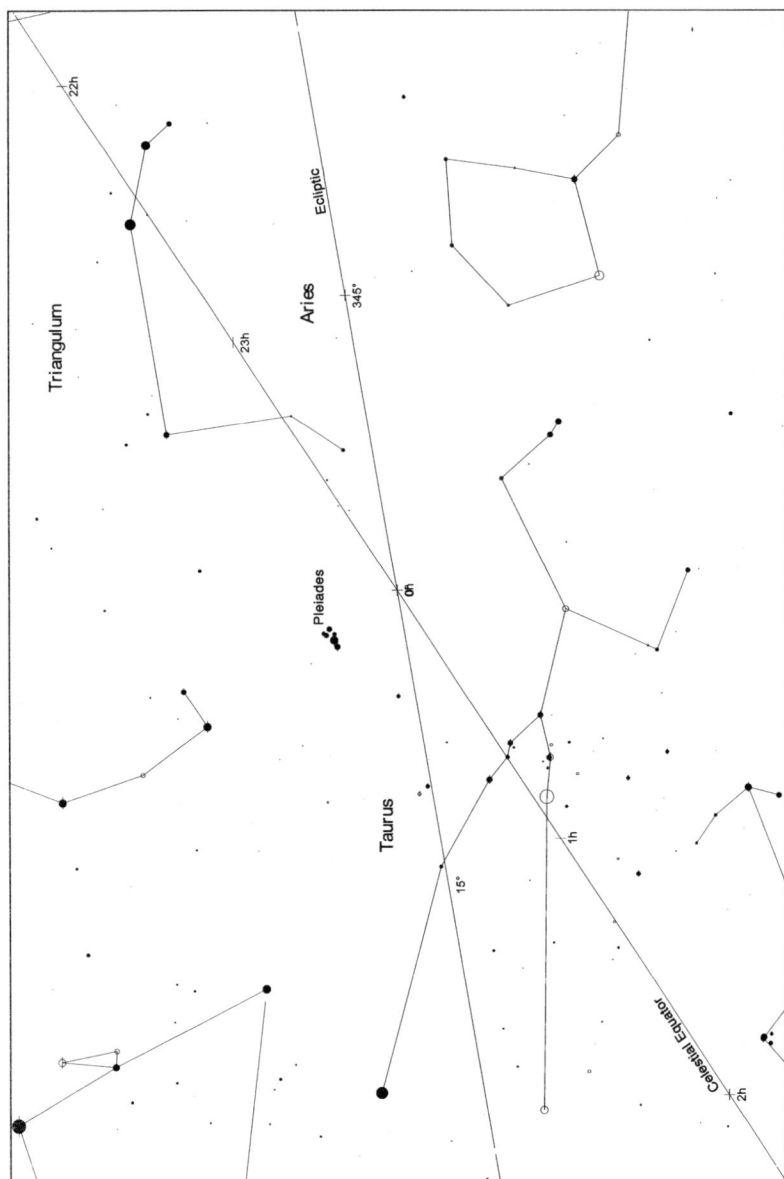

Figure 5.2 The Pleiades
The Pleiades star cluster as a marker of the vernal equinox around 2200 BC (Retrocalculation using Skymap Pro 6).

mentioned in classical sources are a reference, at least in part, to the pre-Celtic people of Britain.

References to the Hyperboreans are found elsewhere in Greek mythology, but the above is the only one that would suggest an association with the British Isles. The classical historians seem to have forgotten who the Hyperboreans were and considered them a legendary race living somewhere in the far north of Europe, "beyond the North Wind". Poets such as Pindar linked them with Homer's Phaeacians or the dwellers in the Islands of the Blessed.[12]

Saturn and the Druids

Pliny tells us that the Druids' calendar was strictly lunar and that their month began on the sixth day after new moon.[13] This is consistent with the notation of the Coligny calendar, which divide the months into a light half and a dark half. Pliny also informs us that they employed the same lunar reckoning to measure longer periods known as 'ages' (*saeculi*) of thirty years; and this interval is also mentioned by Plutarch as being used by the Britons.

A thirty-year cycle has no obvious luni-solar significance, since 30 solar years contain $10957\frac{1}{2}$ days, and thirty lunar years (30 x 354) equal 10620 days. However, it does come close to 371 lunar months, with a small, but significant, divergence from the solar year of roughly 1.4 days. If we assume a lunar year of 12 months, this gives 360 months, leaving 11 intercalary months to be fitted into the calendar; for these to be evenly spaced throughout the calendar they must be inserted about every 2.7 years.

The problem of the half-day could be removed by doubling the cycle to 60 years (i.e. two 'ages' back to back), giving a cycle of 21915 days. This is very close to 742 lunar months with the discrepancy now doubled to 2.8 days, which would then require 22 intercalary months to be fitted into the calendar.[14] This presents difficulties since the extant five-year cycle of the Coligny Calendar clearly shows two intercalary months spaced at $2\frac{1}{2}$-year intervals. If the five-year cycle were simply to be repeated then it would give 12 intercalations in 30 years instead of the required 11. To approximate to a 30-year regime, requires the intercalations to be alternately inserted at $2\frac{1}{2}$ and 3-year intervals. However, we know from the extant fragment that two intercalations spaced at $2\frac{1}{2}$ years followed each other. This suggests the following arrangement for two consecutive 30-year 'ages' of 11 intercalations each:

$$2\frac{1}{2}+2\frac{1}{2}+3+3 +2\frac{1}{2}+2\frac{1}{2} +3+3 +2\frac{1}{2} +2\frac{1}{2}+3 \quad = 30 \text{ years}$$

$$3+ 2\frac{1}{2}+2\frac{1}{2} +3+3 +2\frac{1}{2}+2\frac{1}{2} +3+3 +2\frac{1}{2}+2\frac{1}{2} = 30 \text{ years}$$

Note that if this arrangement were to be repeated after 60 years, then a five-year cycle has to follow another five-year cycle. However, if a simpler eleven-year cycle of five-plus-six years is required to repeat indefinitely then 330 years (30 x 11) must elapse before the calendar repeats. Either arrangement would work (see appendices A & B).

This hypothesis suggests that, in addition to the extant five-year cycle of the Coligny calendar, there should also be a 'lost' cycle of six years. The resulting arrangement is at least as accurate as the Greek *octaeteris*.[15] Certainly, it is greatly superior in accuracy to the old Roman calendar. The full reconstruction given in the Appendices is undoubtedly a functioning calendar and breaks none of the rules of the extant fragment. However, until further evidence comes to light, there can be no proof that the Coligny Calendar actually operated this way.

This still begs the question of *why* the Druids should wish to base a lunar calendar around an imperfect thirty-year 'age'. In this regard, it is instructive to quote the passage by the Greek historian Plutarch, who refers to regular sea voyages, supposedly undertaken by the ancient Britons to a certain mythical island.[16]

> Now when at intervals of thirty years the star of Cronus [*Saturn*], which we call 'Splendant' but they, our author said, call 'Night-watchman', enters the sign of the Bull [*Taurus*], they, having spent a long time in preparation for the sacrifice and the expedition, choose by lot and send forth a sufficient number of envoys in a correspondingly sufficient number of ships.[17]

Therefore, it seems that there is solid evidence that the British Druids did observe the motion of Saturn very closely. If so, then they may have sought some underlying symmetry between the movements of Saturn and those of the Sun and Moon.

It may also be significant that two thirty-year revolutions of Saturn correspond to roughly five Jupiter revolutions of twelve years: the period known as the 'great conjunction'. We have previously encountered this phenomenon when discussing the Chinese sexagenary cycles. It may be that the Druids attached a similar significance to these, the longest astronomical phenomena observable in a human lifetime.

Although Plutarch wrote during Roman times, the fact that the Druids measured the thirty-year orbit of Saturn to and from the constellation Taurus is again significant. It gives us a clue that the astronomy is Neolithic. As we saw in the analysis of the Indian *nakshatras*, before about 2200 BC, the spring equinox occurred in Taurus, and so we have a reason why the earliest Druids chose it as the first constellation of their zodiac.

There is too, a remarkable correspondence between Saturn and the Moon every fifty-six Saturn synods:

56 synodic periods = 56 x 378.09292 = 21173.2 days
717 lunar months = 717 x 29.530589 = 21173.43 days

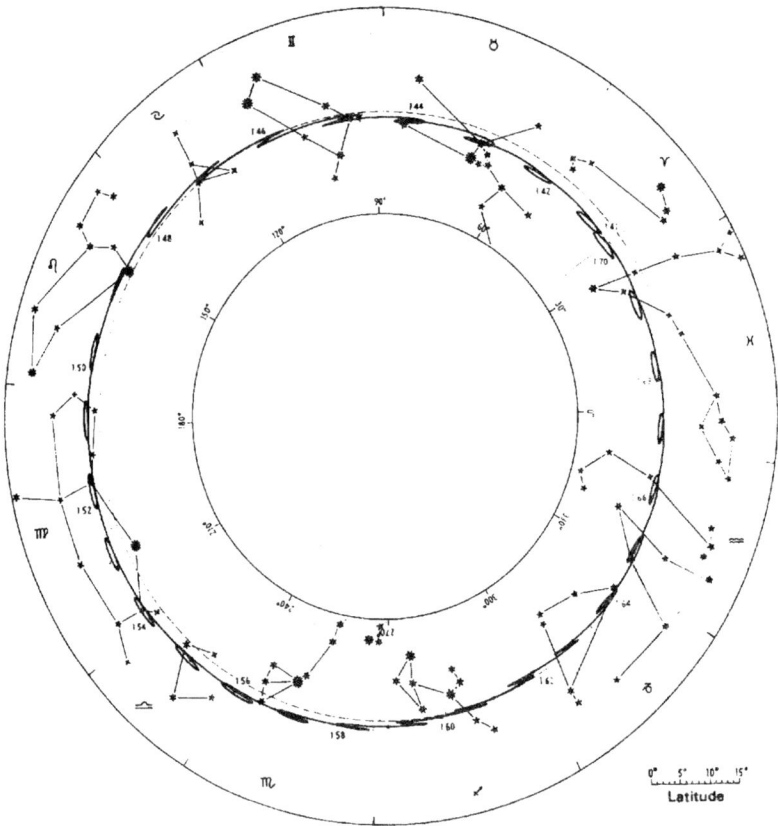

Figure 5.3 The oppositions of Saturn over its thirty-year geocentric orbit.
It seems that the Druids measured the circuit from the constellation of Taurus, back again to the same point; probably using a prominent marker, such as the Pleiades cluster, as their datum. (Reproduced by permission of Floris Books).

The average discrepancy is only about five hours. This means that if Saturn comes to opposition at a certain phase of the moon, then the 56th opposition after this would once again coincide with the same phase of the moon. This is however the *average* period; as the orbits of Earth and Saturn are ellipses, the cycle would have to be observed over many generations to establish the average period. It is an extraordinary coincidence that 56 should be the number of the so-called 'Aubrey Holes' at Stonehenge; a monument of undoubted astronomical significance.

A close solar correspondence occurs after 57 Saturn synods (57 synods = 21551.29 days; 59 solar years = 21549.29 days). The configurations of Saturn are therefore repeated, precisely two days later in the solar year.[18]

Any, or all, of these astronomical coincidences may have been important to the Druids. We should therefore consider the possibility that they utilised the orbit of Saturn as a check on the accuracy of their lunar calendar; and also to provide a long-chronology analogous to those we find in eastern astrology. We have already seen that the synodic period of Venus was employed in just this way within the long cycles of the Mayan calendar. The Old Babylonian astronomers similarly observed the cycles of Venus and considered Saturn the most portentous of all the planets. It would be just as feasible to mesh a calendar to the rhythm of Saturn, if one is prepared to wait a generation to observe the orbit. If we accept that astronomers in Central America, Babylon and elsewhere were performing this kind of astronomy then on what grounds do we deny the same capability to the ancestors of modern Europeans?

We should not forget that our ancestors faced a fundamental problem when they first set out to devise a calendar: how do we *know* that each day, month, or year is the same length of time as the previous one? How do we *know* that time progresses at a constant rate? We may instinctively feel this to be so but how do you set about proving it, with no pre-existing body of experience to call upon? It can only be verified by comparing the days, months and years to some other known datum. It seems that the earliest astrologers chose as a time standard the longest natural cycle available – the orbit of Saturn – and then compared all the other cycles to its rhythm.

This goes some way to explaining the significance of the 30-year Druid 'age' and the postulated 60-year 'double-age'. The equivalence of Saturn and the Moon occurs during the 58th year, but this is not a multiple of either five or six and so there are advantages in basing the calendar around a double 30-year cycle. Consider the following:

60 solar years of 365.2422 days   =   21914.53 days
58 Saturn synods of 378.0928 days   =   21929.38 days
742 lunar months of 29.53059 days   =   21911.70 days

The suggested alternative mechanism would be a simpler eleven-year lunisolar cycle of 4017/4018 days that was allowed to wander, without further correction, alongside a thirty-year cycle based on the observation of Saturn. This 'pure' cycle is actually more reliable than the Metonic cycle over two eleven-year cycles. As may be seen from table 3.1, the lunisolar difference over eleven years is almost exactly a day and a half, so doubling this:

Sun: 2 x 4017.6642   =   8035.3284 days
Moon: 2 x 4016.1601   =   8032.3202 days
Difference   =   3.0082 days

This would mean that the calendar month would slip by about three days in each twenty-two years. Perhaps this explains why the months began on the sixth day of the moon when it was recorded by Pliny. This apart, a lunisolar calendar with three extra days intercalated over twenty-two years will be out by just twelve minutes, whereas the Meton cycle is nearly two hours out each nineteen years (see Appendix B). Thus we see revealed the remarkable accuracy of Plato's Atlantis calendar.

Although we may never know which mechanism was actually used, it may be seen that the three cycles that are actually attested in the classical sources (5-years, 5+6 years and 30 years) are capable of arrangement into a highly reliable calendar. This cannot be a coincidence!

Cronus the Timekeeper

We have surely all seen at some time, the images of Father Time, personified as a bearded old man carrying a sickle and an hourglass. He is the Greek god Cronus and in Greek mythology, his sickle was given to him by Ge (Mother Earth). He used it to castrate his father Uranus (Sky) in revenge for imprisoning the Hecatoncheires. The Greeks named the slowest of the planets after Cronus and the Romans would associate him with their own god Saturn. Indeed our modern words based on the root chrono- are all based upon Greek *khronos* – 'time'. So the very name of the god *Cronus* would have implied 'time' to a Greek.[19]

Apollodorus tells us that many of the Greek myths were of

Hyperborean origin;[20] and Caesar, probably citing Posidonius, would assert that the Celts knew all the same gods as the Greeks and Romans.[21] This may be a further clue to us that the association of the gods with the planets, and of Saturn as a standard of time, have similar origins in ancient Europe.

If we now return to the fragment of Hecataeus quoted above, it is possible that Diodorus was not directly quoting Hecataeus when he discusses the nineteen-year cycle of Meton. It may be that in discussing the great year, Diodorus (or perhaps Hecataeus before him) has merely *assumed* that the nineteen-year cycle of Meton was intended and has interpolated it into the text.

Astronomical considerations reveal that the great year of the Hyperboreans cannot have been the nineteen-year cycle, as this has nothing to do with the period "in which the stars complete their revolutions". It must rather be referring to the orbit of a planet, or planets, around the zodiac.

If instead the passage had mentioned a great-cycle of thirty years, or some multiple thereof, then nothing would prevent us from equating the Boreadae of the Hyperboreans with the early Druids. Festivities took place "from the vernal equinox until the rising of the Pleiades". As the Pleiades star cluster lies in Taurus, then we may equate this to Plutarch's reference that the Britons watched the thirty-year circuit of Saturn as it returned to Taurus. Again, we may recall Pliny, who tells us that the British Druids held religious celebrations "at the end of thirty years".

The reference to the vernal equinox again indicates that the Pleiades cluster must originally have marked the point of the equinox. However, due to the precession of the equinoxes, the drift of the vernal equinox into Aries meant that the Pleiades rose progressively later and later in the year. In Hesiod's Works and Days, a kind of Greek farmers' almanac written about five hundred years before the situation described by Hecataeus, we are told:

> When the Pleiads, daughters of Atlas start to rise
> Begin your harvest; and plough when they go down.[22]

At that era, the Pleiades would have risen before dawn during early May, giving some six weeks of celebrations. Therefore, we may see that the season of festivities must have grown longer and longer until it reached absurd proportions.

The Pleiades would have marked the position of the vernal equinox during the Late Neolithic era before about 2200 BC, when the people of Britain and western Europe were building megalithic stone circles and alignments. This gives us a clue to the era of the calendar and to the identity of the Hyperboreans. It also tells us that they must surely have been aware of the precession of the equinoxes from a very early period.

The Rhythms of Saturn

At this point, it may be appropriate to summarise some of the characteristics of Saturn as they appear to the naked-eye observer.

Saturn shares most of its characteristics with the other superior planets, Jupiter and Mars, the main difference being its longer period and hence, slower motion. Venus and the elusive Mercury, which never stray far from the vicinity of the Sun, can be seen only during the morning or evening hours. However, the superior planets can be observed in a dark sky throughout the night.

When the planet stands in *opposition* to the Sun it will rise in the eastern sky just as the Sun sets in the west. It will then reach its highest point in the sky (its culmination) exactly at midnight. As the year progresses, it will rise progressively closer to the sun until it becomes invisible in the evening haze. During the period of *conjunction* with the Sun, the planet is not visible, until it appears again in the haze of the morning sky. This first sighting after conjunction is termed the *heliacal rising* and its disappearance in the evening haze is the *heliacal setting*. The period between the recurrences of similar events is its *synodic period*.

In the case of Saturn, the average synodic period is 378.09 days, or just under 13 days longer than a solar year. This means that Saturn will return to opposition 13 days later than in the previous year. For most of the year, the planet appears to move forward in the zodiac (i.e. in the same direction as the Sun) but once in each synodic period, it performs a retrograde loop. The planet reaches a *stationary point* after which it moves backward relative to the stars for about 140 days and then after a second stationary point it begins to move forward again. The opposition occurs roughly in the middle of this retrograde loop.

The long-term rhythms therefore recur in multiples of a year and thirteen days, and so each opposition progresses round the zodiac, until after roughly thirty years (the sidereal revolution) have passed, it returns again to the same constellation of the zodiac. However, as 28 synodic periods are slightly shorter than 29 years this means that the retrograde loops are displaced by 5° 18' to the west of the previous revolution.

The period of the sidereal revolution is 10759.2 days or 29.458 years. However, this period would be difficult to measure directly because of the retrograde loops. It would be much easier for naked-eye astronomers to measure the synodic arc, between successive stationary points. We may note that:

$$378.09 \times 29 = 10964.61 \text{ days}$$
$$10964.61 \div 365.2422 = 30.02 \text{ solar years}$$

The discrepancy of 7.3 days is almost a quarter moon, or one seven-day week. We may therefore see that it would be possible to schedule a thirty-year Saturn festival a quarter-moon later each thirty-year cycle – provided that the calendar is properly adjusted to the solar year by intercalation. If the calendar cycle is adjusted to solar time to keep the seasons in step then the lunar accuracy has to be sacrificed.

Again, it must be stressed that these can be only suggestions! Just because a calendar could operate on these principles is no proof that it was actually used that way, and any number of reconstructions are possible based on the extant Coligny fragment.

The Antediluvian Calendar

We have then, several clues that that a north European race known as the Hyperboreans were performing complex astronomy as early as the third millennium BC. We also have archaeological evidence from Britain and Gaul of the building of stone circles and astronomical alignments between about 3000 BC and 1500 BC. It is then a small step to equate the two. With this in mind, it is also instructive to look again at the various Mediterranean myths of a civilisation on an Atlantic island, which were discussed in Chapter One; and consider that these too may recall the British Isles during Neolithic and Bronze Age times.

We began this chapter with a discussion of the reference in Plato's Atlantis myth, to an 'antediluvian' calendar with alternating five and six-year cycles. It has been demonstrated above that this calendar actually works. If, as has been suggested previously, the antediluvian year contained only 360-days then this calendar would also have to work under such a regime.

The Sun, Moon and planets are supremely indifferent to any cataclysm that might befall the Earth. If the Earth suffered a devastating comet impact then they would go on as before, with their periods unaffected by any change to the length of day here on Earth. Only the Moon might be tidally affected (since angular momentum lost by the Earth must be gained by the Moon) but this is unlikely to have been noticed. Therefore, we may establish the former calendar rules by simple ratio.

$$\text{One old-day} = 365.25/360 \times 24 = 24.35 \text{ hours}$$
$$\text{One old-lunar month} = 29.53 \times 360/365.25 = 29.1 \text{ old-days}$$
$$\text{One old-lunar year} = 12 \times 29.1 = 349.2 \text{ old-days}$$

There is little point in trying to be more accurate than this. It is unlikely that the antediluvian solar year was precisely 360 days long, any more than it is now exactly 365 days. However, we may expect it to have been close to this value. We may see that the relationship between Sun, Moon and planets is unchanged, merely that each cycle contains fewer days.

We may also deduce the former number of days in thirty years by the same ratio.

$$10957.5 \times 360/365.25 = 10,800 \text{ old-days}$$
$$30 \times 360 = 10,800 \text{ old-days}$$

We have met this number before. It is one of the important Brahmanic numbers of the Indian calendar cycle and 10,800 *years* was considered by Heraclitus to be the period of the Great Year. This is surely another extraordinary coincidence.

We may now see what the significance of this number was. Not only was it the number of days in thirty years, it also represented the revolution of Saturn.[23] Each orbit symbolized a divine day of the gods. Therefore, a divine *year* of the gods was 360 revolutions of Saturn or 10,800 *years*. The symmetry of heaven reflected the reality on Earth. The cataclysm of the Great Flood destroyed this beautiful symmetry. It no longer works. However, in India and Babylon, it was never forgotten and has come down to us shrouded in eastern religious imagery. The preservation of this divine number, together with the evidence for a 360-day year in the earliest calendars, is further corroborative evidence that the Flood was an ancient reality.

Circles and Standing Stones

The association of the legendary Hyperboreans with the megalith builders of Britain and Atlantic Europe is a far from new suggestion; and one might even venture to say that it is obvious. The mythology of the megalith builders is lost to us, except perhaps for that which may be preserved among the Celtic myths.

A reading of the fragment of Hecataeus again reminds us that all ancient astronomy was really astrology; and that the builders of astronomically aligned monuments were only concerned with scheduling the particular appearances of their deities for ritual purposes. Any pure astronomy that they learned was a by-product of this.

In the land of the Hyperboreans, there was a precinct sacred to the sun god Apollo and nearby, a "spherical temple" supervised by the Boreadae. Many theorists have therefore identified this spherical temple with Stonehenge. Even if one of the other British stone circles is intended then Stonehenge must surely have been even more magnificent. However, many of the astronomical alignments claimed for these monuments by earlier theorists are now disputed.[24] It seems that the monument we see at Stonehenge today was a temple embodying only symbolic orientation, but earlier phases of the monument may show evidence of more practical alignments. One early investigator even thought that he could see evidence to link Stonehenge with Saturn astronomy.[25]

The first phase of Stonehenge dates from the early third millennium BC and was contemporary with other great circles in Britain, such as those at nearby Avebury, Castle Rigg in Cumbria and the Ring of Brodgar on the Orkney Islands. It consisted of the circular bank and ditch and the entrance causeway oriented towards midsummer sunrise. Within the enclosure, the fifty-six holes that we now name after the historian John Aubrey were dug – and shortly afterwards refilled with chalk. They may once have held wooden posts. The Heel stone on the causeway and the four Station Stone positions also date from this early phase. The great Sarsen circle that now stands at the centre of the monument was not begun until about 2100 BC.

In summary, the various theories concerning alignments at Stonehenge (for those at other monuments are very similar) are primarily concerned with marking the extreme rising and setting points of the Sun and Moon on the horizon. The return of the Sun to a particular position marks the passage of a year. As such, it is an alternative method to establish the number of days in the year, rather than measuring the length of the solstice shadow as the Chinese were doing at this same era.

The motion of the Moon is more complex, because its orbit is tilted at an angle of 5° 9′ to the ecliptic and is subject to precession, which pulls the entire orbit round in a circle every 18.61 years. This means that its rising and setting positions on the horizon vary, with this periodicity, between extremes that lie about 10° either side of the Sun's extremes. These variations are of little significance if all you wish to do is construct a calendar. They only assume an importance if your purpose is to predict *eclipses*. These occur only when both Sun and Moon are near one of the *nodes* where the orbit of the Moon crosses the ecliptic.

Theories of Stonehenge as an eclipse predictor were proposed in the 1960's by Hawkins and a decade later by Hoyle, with slightly different emphases. Hawkins thought that the 19-year cycle of the Hyperboreans, as described by Hecataeus, was really an allusion to the 18.61-year cycle of the Moon's nodes.[26] In fact, the two concepts are quite unrelated. He also pointed out that a cycle of 56 years (19 + 19 + 18) was closer to the true period of the nodal-cycle; and that the 56 Aubrey Holes could therefore be used as an eclipse computer.

Hawkins suggested that by using six stones spaced 9, 9, 10, 9, 9, 10 holes apart, rolled one hole counterclockwise each year, could be used to predict all the important lunar phenomena. Hoyle's method was more complex, but with the same objective. Two stones are rolled around the circle such that they track the regression of the two nodes around the ecliptic. Other stones followed the position of the Sun and Moon. When the Sun and Moon markers are close to opposite nodal markers then an eclipse of the Moon is possible at full moon; when Sun and Moon markers both approach the same node then an eclipse of the Sun is possible at new moon. Whether the eclipse will actually be visible from that location is beyond the accuracy of the method.

Few researchers would now support theories that Stonehenge was an 'eclipse computer'.[27] The power of the mainframe computers that Hawkins used to calculate the ancient positions of the Sun and Moon are dwarfed by that of any modern home computer; and commercially available software will now enable anyone to recreate the conditions of the Neolithic sky.

The work of C. A. Newham is less well known, but his investigations into astronomical alignments at Stonehenge were both contemporary with and completely independent of Hawkins and Hoyle.[28] Although equally scathing of the sceptical archaeologists, he remained unconvinced that Neolithic alignments were concerned with predicting eclipses, saying:

> ...there is no evidence of such an intention. It would certainly have been more difficult to define eclipse cycles by this crude method.[29]

The equating of the nineteen-year cycle of the Hyperboreans with the 18.61-year node cycle is tenuous; and as was discussed above, the reference to the nineteen-year cycle by Hecataeus may itself be suspect. Perhaps the most obvious objection was given by Hoyle himself. The same eclipse predictions could be achieved by recording the positions on a peg-board – there is no need to build a huge outdoor circle at all.[30] Furthermore, after a century or so of recorded observations, the periodicity known as the *Saros* would become apparent and the observations then become superfluous.[31] Other societies, in China, Babylon and the Americas all succeeded in predicting eclipses this way, without recourse to megalithic observatories.

The stone circles were more likely ritual centres possessing only symbolic orientation. The ceremonies were intended to be seen by the assembled population and the astrological portents with which they were associated could be demonstrated to all just by looking up at the sky.

The few classical references that we have are sufficient to tell us what Celtic and Hyperborean rituals were really about. These, together with the Coligny Calendar, suggest that festivals and sacrifices in honour of the Sun god were held according to a strict lunisolar cycle. However, the most important festival of all was held only once each generation and was associated with the return of Saturn to the Pleiades (or perhaps to the vernal equinox) at roughly thirty-year intervals. The precise date of the festival had to be scheduled in the calendar and determined by direct observation. Few people would experience more than one Saturn festival during their lives and so the alignments had to be set out in immutable stone to be remembered from generation to generation.

The number that links Saturn and the Moon is fifty-six.[32] After 717 lunar months and 56 synodic revolutions of Saturn there is an equivalence that would allow the calendar to be checked and adjusted. The festivals of the next long-cycle could then be set. If we can accept that the megalith builders were prepared to wait fifty-six years to follow the eclipse cycle then we may equally accept that they would wait thirty or sixty years for Saturn. We find clear references in written classical sources that the ancient Britons concerned themselves with Saturn. There are no references whatsoever from any classical source to indicate that they were interested in eclipses.

Again, we come back to the question: *why?* Why were they so interested in the portents of Saturn? Why expend generations of effort in monument building? The answer is in part supplied by the evidence we have seen from other ancient civilisations. The planets were gods and the gods determined the fate of all things on Earth. Certain planetary alignments and conjunctions would signify the end of the world and perhaps the gods could be propitiated by timely rites and sacrifices.

We cannot know whether the people of Neolithic Britain and Europe had a calendar era like those of India and Mexico. We have already seen indicators that there was a flood catastrophe at some time around 3100 BC. So for the stone circle builders of 3000–1500 BC the Flood was not just a myth as it is to us, but a historical reality.

One consequence logically follows another. If one accepts this date for the Flood then the world may still have been wobbling on its axis throughout this entire period; and the length of the day might not be a constant either. For Chinese astronomers measuring the number of days in the year by the length of the shadow, then the solstice shadow would appear to oscillate in length. For the horizon astronomers of Europe, the Sun and Moon would appear to rise and set to the north and south of their average positions over the period of the wobble. For calendar-makers the number of days in the year – and in longer cycles – might not be reliable. Calendars based solely on day–counts and cycles would soon drift into error. Only the planets kept proper time. So does this perhaps tell us why ancient people were so obsessed with planetary observation and astrology?

However, we have touched upon the subject of eclipses; and eclipses offer us another window on ancient chronology. We must now move on to see if these can tell us anything about the subject of catastrophism in human prehistory.

References to Chaper Five

1 Plato, Critias, 119.

2 See the appendix to my earlier book *The Atlantis Researches.*

3 W.F. Kirby (London 1907) *Kalevala: the Land of Heroes,* (republished by the Athlone Press, 1985); see Runo L, p 634.

4 Caesar, The Gallic War, VI.xiii–xiv; VI.xviii.

5 Caesar, The Gallic War, VI,Xiii

6 Duval, P.M. & Pinault, G., 1986

7 MacNeill E, (Dublin 1928), On the Notation and Calligraphy of the Calendar of Coligny, Eriu, X, (1926-8), VI.

8 For example, Fotheringham and Rhys. These papers are critically analysed in the work of MacNeill and Olmsted.

9 Olmsted (1992) has suggested that the Coligny calendar was based upon a theoretical 25-year cycle, but agrees that this replaced an earlier 30-year cycle. However, while there is evidence for a five-year festival, there is no textual evidence for the existence of a 25- year cycle in ancient Gaul.

10 Diodorus Siculus, II, 46, 47; translation by E. Davies, 1804

11 Hawkins, G.S, 1966

12 Pindar, Olympian Odes 3.16; Pythian Odes 10.39-10.45

13 Pliny, Natural History, XXX.xiii

14 This was in fact the solution that I suggested in *The Atlantis Researches.* Although this would work, there is no textual evidence for the use of a 60-year cycle. I therefore now favour a simpler rolling 11-year lunar cycle, of five years plus six years, repeated indefinitely.

15 Geminus, Elementa Astronomiae, C.8.

16 On the subject of pre-Celtic British voyages to America, see the interesting ideas in *The Alban Quest* by Farley Mowat.

17 Plutarch, The Face in the Moon, 941. From the translation by H. Cherniss and W. Helmbold; Loeb Classical Library

18 There is evidence that the Babylonians observed the synodic arcs of the planets and obtained a figure for Saturn of 57 revolutions in 59 years: 365.25 x 59/57 = 378.06 days. On this, see Pomeras & Taton, 1963, pp118-120.

19 Plutarch, Isis and Osiris, 326,32

20 Apollodorus, 11, v, ii ;

21 Caesar, The Gallic War, 17, 1-2

22 Works and Days, 383-384; from the translation by D. Wender

23 Actually, the correspondence is still about a week out, but ancient people may have believed the correspondence to be exact. Note that: 10964.61 × 360/365.25 = 10,807.007.

24 See North, J.,1986, pp 407. He suggests:"The idea was that – in ways unexplained – the relation between this [lunar] cycle and the cycle of abnormal tides of the sea had been discovered, and that Stonehenge would thus have been capable of facilitating cultural communication across the North Sea and English Channel". Hopefully, this explanation has been addressed here.

25 Gidley, L., 1873.

26 Hawkins, G, 1965, pp 174-177

27 North, J., 1996, pp 406-8

28 Newham, C.A. 1974 and 1970

29 Newham, C.A. 1972, p 20

30 Hoyle, F, 1977, p 80

31 Hoyle, F., 1977, pp 100–101

32 Plutarch (Isis and Osiris, 362) associates the number 56 with the daemonic power of Typhon (Egyptian Set) and the number 12 with Zeus. Twelve may be the passage of Zeus (Jupiter) though the constellations of the Zodiac in each orbit – but what is the significance of the number fifty-six?

6

Disposing of Eclipses

The work of generations of historians enables us to assert a precise date for many of the events of classical history. This certainty is accomplished because the Julian calendar succeeded the old Roman calendar, and the calendar dates of adjacent nations can usually be converted to Roman dates, so long as at least one event is specified in both calendars. Once a single date is known then, so long as we know the intercalation rules, we can convert any calendar date to the Julian calendar.

Greek dates can usually be dated by reference to the year of the Olympiad, which took place in octaeteris and half-octaeteris years. Babylonian dates can be correlated using Ptolemy's astronomical almanac, based on the Egyptian wandering year, which goes back to the era of King Nabonassar. His reign began on 27 February 747 BC, although not all the subsequent reigns can be stated with such certainty. For Egypt, we rely on Manetho's list of kings. These give us a firm date of 525 BC for the Persian invasion by Cambyses, from which earlier reigns can be counted back. In fact, Egyptologists are confident of the reigns back to 664 BC when the Assyrian king Ashurbanipal attacked the Egyptian city of Thebes – an event recorded by both Berossus and Manetho.

Before these dates, the firm links to the Julian calendar are lost. Historians therefore have to rely upon counting back the reigns found in the various king lists, which are subject to human errors both in transmission and in recording. In Manetho's Egyptian king list, for example, we find dynasties of kings who reigned in parallel in different parts of the country. Other kings were omitted from the lists as heretics or usurpers. Although specialists can fix dates for some kings, by reference to calendar dates in inscriptions, this cannot confirm the arrangement of the reigns between the fixed points. Moreover, archaeologists know of a number of king lists, which do not always agree with each other. History for these early periods is reduced to the experts' best opinion of the past and whether or not we believe the experts.

Astronomical Certainties

Historians like to find references to astronomical events in ancient records. These can be retro-calculated and if they can be tied to a year of a king's reign then they provide an independent fixed point from which reigns can be counted. Of particular value here, are records of ancient eclipses, as these can be back calculated to a precise day and hour.

A reference to any suitable astronomical event may be used by historians to fix a date, or range of dates. Egyptian chronology was founded upon a series of Sothic dates: references to the Egyptian civil calendar within inscriptions, which can be associated with the regnal year of a king. For Mesopotamia, it is observations of Venus that have provided a way into the king lists. These dates continue to be the subject of controversy.

For geophysicists the problem is somewhat different. They would like to investigate the irregularities in the Earth's rotation as far back as possible, to see how it has changed. Now it must be stressed that there is nothing catastrophist about such enquiries. The Earth's rotation is gradually decelerating due to the tidal drag of the Moon on the oceans and this has gone on for millions of years, since the formation of the Earth-Moon system. The length of the day is also slightly irregular. Although the range of variation is only in the order of milliseconds per year, this accumulates to a significant discrepancy in the timing of ancient astronomical sightings.

The tidal retardation of the diurnal rotation was first recognised by Immanuel Kant in 1754. In 1781 astronomer William Herschel pointed out that no-one had ever investigated whether the Earth's rate of rotation was constant; and that this was somewhat difficult since the day was itself the time standard by which other periods were measured.[1]

Modern astronomers have therefore defined an independent time standard based on atomic clocks, called Terrestrial Dynamical Time (TDT) which is used in astronomical calculations. A standard reference day is defined as 86,400 SI seconds. Time on Earth however, is still measured in Universal Time (UT) which was formerly called Greenwich Mean Time. This is based on the day from midnight to midnight at the Prime Meridian.

If the diurnal rotation were constant then the location reported for historical eclipses would always agree with the calendar date and time expected. However, the irregular tidal retardation of the day means that the eclipse track will fall further east than expected. Astronomers refer to this small discrepancy (TDT - UT) as ΔT (delta-T). The cumulative discrepancy between clock and calendar due to this slowing has been calculated at about half a day since 1500 BC.[2] Therefore, any astronomical event of that era that we would expect to be visible from a place on the Earth's surface would actually have been seen on the opposite side of the world.

We may therefore see that this is another problem that chases its own tail. The geophysicist wants a precise astronomical observation that can be assigned by non-astronomical means to a calendar date and time of day. Historians on the other hand, are satisfied with a date from the astronomers that will fix a historical event to a calendar year.

The Mechanism of Eclipses

Before proceeding further, some introduction to eclipse science is required, for the benefit of non-astronomers.

An eclipse of the Sun can only occur at new moon when the Moon passes between the Earth and the Sun and casts its shadow on the Earth. A lunar eclipse can only occur at full moon, when the Moon passes into the Earth's shadow. The reason that eclipses do not occur every month is that the Moon's orbit is tilted with respect to the geocentric orbit of the Sun (the ecliptic). The two points on the celestial sphere where the orbital planes intersect are called the *nodes*. These are termed the *ascending node* where the Moon crosses from south to north of the ecliptic; and the *descending node* where it crosses north to south. A solar eclipse occurs when Sun and Moon align near the same node; a lunar eclipse occurs when they align near opposite nodes.

As was discussed in the previous chapter, the orbit of the Moon also precesses such that the nodes regress round the orbit in an 18.61-year cycle.[3] The alignment of the Sun and Moon recurs in a period 18.62 days less than a solar year, or 346.62 days. This is termed an *eclipse year*. However, as there are two nodes, an eclipse of some kind is possible every 173.31 days. A month in which such an alignment falls is known as an *eclipse season*. Eclipses may be only *partial*, if the alignment is not perfect.

Other factors come into play. Both Sun and Moon appear nearly the same size in our sky, but as the Moon's orbit is elliptical, its distance and apparent diameter varies significantly. For lunar eclipses, this makes little difference to the appearance, but for a solar eclipse, it means that they vary in magnitude. A maximum *total eclipse* may last over seven minutes when the Moon is at perigee (its closest approach). If the eclipse occurs near apogee then an *annular eclipse* occurs, where the umbra of the Moon's shadow doesn't quite reach the Earth's surface. In an annular eclipse, a ring of sunlight remains visible around the Moon.

Most people should observe a number of lunar eclipses in their lifetime, but few except committed 'eclipse chasers' will have seen a solar eclipse. This is because each lunar eclipse can be observed over the entire night-time hemisphere, whereas the path of the Moon's shadow touches only a small area of the Earth's surface. On average, any particular location should experience a solar eclipse about every 375 years.[4] Sometimes the path of totality does not cross any inhabited land and many solar eclipses, ancient and more recent, must have been viewed only by fishes and penguins.

Although the eclipse season tells us when an eclipse should occur, somewhere in the world, it does not tell us whether it will be visible from our location. Since ancient people could not chase eclipses around the world, they could only record those that occurred in their neighbourhood; and in those cultures where records of eclipses were kept, a periodicity became apparent.[5] The Babylonians noticed the following relationship:

$$223 \text{ synodic months} = 223 \times 29.53059 = 6585.32 \text{ solar days}$$
$$= 18.03 \text{ years}$$
$$19 \text{ eclipse years} = 19 \times 346.62 = 6585.78 \text{ solar days}$$

The difference between the two cycles is just over eleven hours and the period of 18 years and 11 days is known as the eclipse *Saros* period. Eclipses linked by this relationship are said to belong to the same Saros series.

The name Saros however, has nothing to do with the Babylonian saros of 3600 years previously mentioned. It is a misapplication of the term by modern astronomers, dating back to Edmund Halley in the eighteenth century.[6] The German astronomer Theodor von Oppolzer employed a team of assistants to calculate the dates of eclipses back to

1207 BC and his Canon of Eclipses was posthumously published in 1887. He numbered each Saros and each eclipse within it, and this system remains in use by modern astronomers.

The Saros interval is not a whole number of days. The difference of 0.32 of a day is just under eight hours, which means that every third eclipse in a Saros series (3 x 18 = 54 years) should be observable from the same longitude. In fact, it occurs about 10° west of that of three Saros earlier. Moreover, the 11-hour discrepancy between the two cycles causes a progressive change in latitude of the eclipse path, north or south of the previous one, depending whether the eclipse is at the ascending or the descending node.

Another factor is the *anomalistic month,* the period in which the moon returns to the perigee of its orbit. As this period is 27.555 days, it may be seen that there is a further correspondence:

$$239 \times 27.5546 = 6585.5494 \text{ days}$$

This means that solar eclipses within a Saros series will have similar characteristics. From the viewpoint of an ancient observer, if the previous eclipse were total, then the next eclipse would also be total. If it were annular, then the next would be annular. This is why the Saros series was so useful for astrological predictions.

Lunar eclipses are somewhat simpler, as every third lunar eclipse (3 x 18.03) will certainly be visible, only slightly earlier in the evening; and one of the intervening eclipses in the series may also be visible from the same location. It is perhaps more likely that ancient astronomers first noticed a periodicity in the occurrence of lunar eclipses and could then warn that a solar eclipse was *possible* at either the preceding or the following new moon.

Eclipse Demons

The cause of eclipses was not apparent to ancient people. There are numerous myths of demons and monsters consuming the Sun and Moon. In India, this was Asura, the sunlight-demon, who intervened to prevent the Sun's light from reaching the Earth. In Norse mythology, it was a great wolf that chased the Sun and Moon around the sky and occasionally caught up with them. The earliest Chinese name for eclipses was *shih,* 'to eat' implying that the Sun and Moon were being eaten by a dragon.

Many other strange ideas circulated. Heraclitus, who was a philosopher not an astronomer, believed that the Sun and Moon were bowl shaped; and that eclipses occurred when their concave inner surface was turned towards the Earth.[7]

In the west, the discovery that eclipses were caused by the shadow of the Earth or Moon is attributed to the Athenian Anaxagoras (500-428 BC). Plutarch tells us that his pupil Pericles allayed the fears of the sailors when an eclipse occurred during a sea battle of the Pelopponese War. He held his cloak above his head and mocked the fearful sailors:

> What is the difference, then, between this and the eclipse, except that the eclipse has been caused by something bigger than my cloak?[8]

The Chinese certainly understood the true cause of eclipses by about AD 80. The Han astronomers used a prediction cycle of 135 lunar months (11½ eclipse years) rather than the Saros – a circumstance incidentally, which argues against Babylonian influence in other areas of their astronomy.[9] However, a purely astrological interest in the prediction of eclipses does not require the cause to be understood. It merely requires observations to be religiously recorded against an *accurate* calendar.

The many myths of a dragon devouring the Sun during an eclipse may recall the sighting of a comet close to the solar disc during an eclipse. It is also quite likely that, by pure chance, an eclipse must have occurred while a great comet was visible in the sky. It may be that this also gave rise to many of the myths of second or multiple suns in the sky.

While tales of eclipse demons may have held a place in popular folklore, we should beware of glib statements that ancient astronomers did not understand the cause of eclipses. We need not doubt that the true cause was recognised by many ancient astronomers, in many places, but was forgotten again. The court astronomers of Babylon, who were charged to watch for the first thin crescent of the Moon, must surely have observed partial and total eclipses. Can we seriously accept that they did not know that it was the Moon, which had passed in front of the Sun? It is all part of the textbook conspiracy that insists *ancient* astronomy was *bad* astronomy.

Early Historical Eclipses

The Alexandrian astronomer Claudius Ptolemy in his *Almagest* (c AD 150), cites a table of Babylonian eclipse records and planetary conjunctions, most of which were drawn from the earlier work of Hipparchus. Although these date back only to 700 BC, other records exist in the form of excavated cuneiform tablets. One such eclipse recorded in the Assyrian *Limmu* lists, identified as that of June 15 763 BC, has enabled scholars to fix the reigns in the Assyrian king lists back beyond 1100 BC.[10]

Among the cuneiform records are so-called "goal-year" texts. These give lists of eclipses separated by eighteen years (the Saros interval). Some of these extend over several centuries and show the predicted eclipses for the goal year, based on historical eclipses. We may presume that it was something like these tables that Alexander's astronomer, Callisthenes, sent back to Greece from Babylon in 331 BC.

An even older source of observations comes from the Chinese oracle bone divinations, which go back to the Shang Dynasty around 1500 BC. A few other scattered references exist in literary sources, which may be of interest to the historian, but none supply the precise timings that a geophysicist needs. They are all potentially of value to the student of catastrophism.

The most detailed of these ancient records, together with more recent data from Europe and Arabia, have been investigated by Stephenson and Houlden. Their conclusions have enabled them to calculate the value of ΔT to a high degree of precision back to about 700 BC.[11] Earlier than this, the value can be extrapolated using the expressions that they derived; and most retro-calculation computer software now utilises such formulae.[12] The effect of ΔT is to produce a displacement in the longitude of ancient eclipse paths, to the east of that expected if the length of day were constant. The latitude is unaffected.

The rate at which the diurnal rotation is slowing appears to have changed at some time around AD 700-1000.[13] For the past 1,000 years, the rate of change is about 1.4 milliseconds/century whereas before this the rate was nearer 2.4 ms/Cy. The reason for this change is not known and leads to a conclusion that either the earlier data is unreliable, or some other non-tidal force was at work.[14]

An example of a historical event that can be dated by astronomy is the eclipse described by Herodotus, which occurred during a battle between the Lydians and the Medes.[15] He claims that Thales of Miletus had predicted the eclipse, although the method he employed is not stated. After many years of stalemate the warring parties took the eclipse as a portent and were eager to make peace.

Pliny credited Thales with discovering the cause of eclipses and supplies the date of the eclipse as 585-584 BC by reference to the Greek Olympiads.[16] The date of the battle can therefore be placed on *May 28, 585 BC*. Pliny also tells us that this was year 170 after the foundation of Rome. So if we had no other means of doing so, then we could also date the legendary foundation of Rome to 755 BC by means of this eclipse.

The eclipse seen by Pericles during the Pelopponese War can be equated with the annular eclipse of *August 3 431 BC*, from the information given by Thucydides in his history of that war.[17] Thucydides may have seen the eclipse himself. He says that the Sun was reduced to a thin crescent and stars were seen. Modern calculations suggest the eclipse was only partial at Greece and so the darkening of the sky is difficult to explain.[18] Could it imply a difference in the latitude of the path?

In recent decades, the oldest record of an eclipse was claimed to be that recorded on a clay tablet from the palace archive of the ancient Syrian city of Ugarit. Archaeologists dated it loosely to the late Bronze Age, after which the city was abandoned. As few Ugaritic texts are known, the specialists have disputed the meaning of the text, but one translation of it reads as follows:

> On the *btt* day of the new moon in (the month) *hiyaru* the Sun went down, its gate-keeper was *Rsp*. Two livers were examined: danger[19]

The meaning of *Rsp* (Rashap) was interpreted, by the same specialist, as the planet Mars and the term *btt* as the sixth hour of the new moon. The month of *hiyaru* is here interpreted as falling around late February or early March, based on an assumption that the Ugaritic calendar was similar to the Egyptian lunar calendar. However, other investigators have declared that the text merely refers to the presence of Mars during the first six days of the new moon.[20]

Another opinion of this troublesome text gives it as:

> The day of the new moon of Hiyaru was put to shame as the sun set, with Rashap as her gate-keeper[21]

On this translation, an eclipse occurred at sunset – a very rare event indeed, and if correct, it would conclusively identify the date. The identification of Rashap with Mars is also disputed. Again it comes down to the opinions of the specialists and which one we choose to believe.

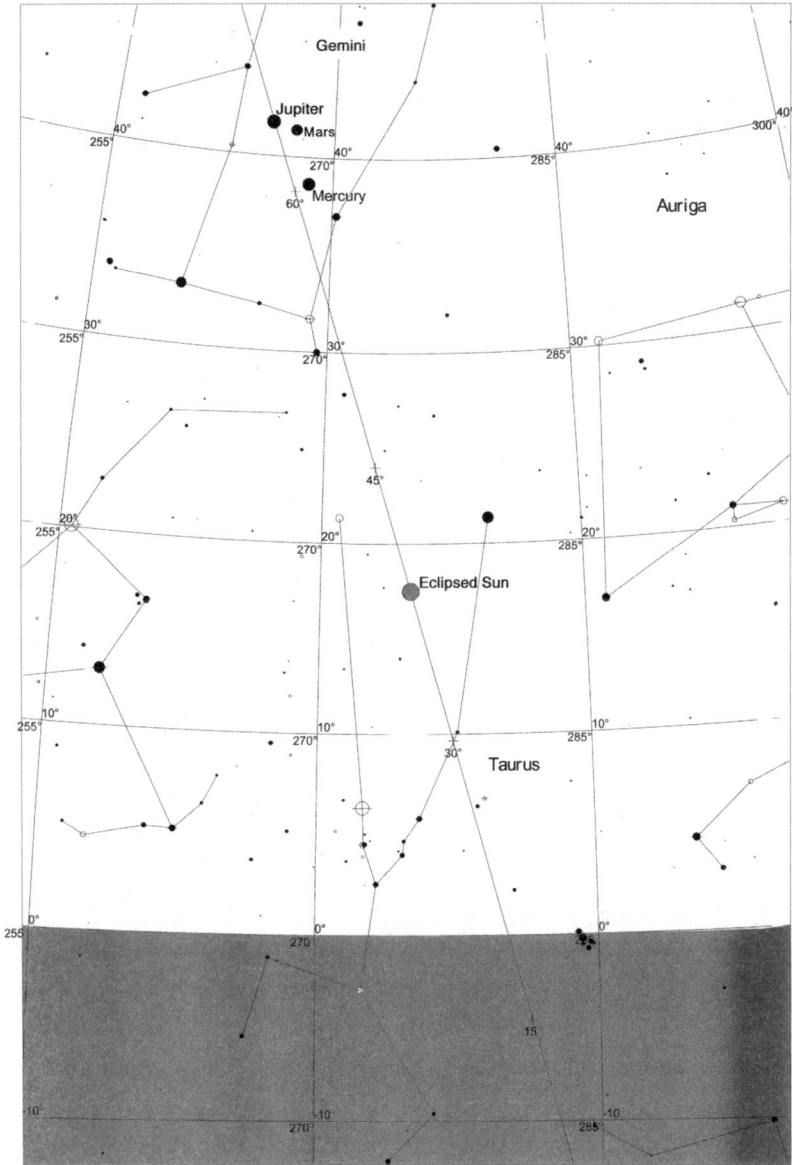

Figure 6.1 Eclipse seen at Ancient Ugarit
The unique circumstances of a sunset eclipse recorded at ancient Ugarit (Syrian coast) on 9 May
1012 BC. The retrocalculation shows an eclipse in the horns of Taurus, with a triple conjunction
of Jupiter, Mars and Mercury above in Gemini. Although this is in close agreement with most
translations of the Ugarit text, the retrocalculation shows that the track passed over southern
Turkey and was not total at Ugarit. Clearly there is much still to be learned from this eclipse.
(Retrocalculation using Skymap Pro 6).

The claim that this was the oldest positively dateable eclipse record goes back to the arguments of Sawyer and Stephenson, who identified it with the eclipse of *3 May 1375 BC*. This was based mainly on their identification of *Hiyaru* with Babylonian *Ajjaru*, which was a spring month.[22] De Jong and van Soldt then disputed this and suggested *5 March 1223 BC*, on the basis that this was the only eclipse during which Mars was above the horizon. However, a reconstruction of these eclipses, based on the currently accepted values for ΔT, will not permit their paths to cross ancient Ugarit.

The solution proposed by Mitchell was a lower date of *9 May 1012 BC*, which can indeed be reproduced (see figure 6.1). The circumstances show an eclipse at sunset, with the Sun in the horns of Taurus – indeed the Sun set while still in eclipse. Above the eclipse is a three-planet conjunction of Jupiter, Mars and Mercury. It would be an extraordinary coincidence if a mistranslation of an ancient text were to give the circumstances of such a unique eclipse. One anomaly however, is that the calculated path of totality does not pass directly over Ugarit and would have to have been viewed from further north, in southern Turkey. A small error in ΔT would make little difference in this instance, as the track is almost west to east.

Egyptologist David Rohl has attempted to use this eclipse (along with much other evidence) to propose a revision of Egyptian chronology for the New Kingdom and Third Intermediate Period.[23] The burning of the palace at Ugarit, where the clay tablet was found, is mentioned in one of the Amarna letters written to the heretic pharaoh Akhenaten. Based on this, Rohl would revise the date of Akhenaten's reign to the late eleventh century BC, whereas the conventional chronology of Egypt places his reign in the mid-fourteenth century BC. This would then have a knock-on effect for all the earlier chronology of Egypt and the surrounding regions. We therefore see the importance that just one conclusively identified eclipse could have for our understanding of ancient history.

The claim for 'oldest solar eclipse' therefore passes to a Chinese oracle bone. Many of these inscriptions were used by Stephenson in his computation of ΔT and the best candidate seems to be a divination from Anyang. The bones were used rather like a modern reading of tealeaves! The diviner would make cracks in a piece of bone using a hot needle and then, together with other available portents, attempt to foretell future events based on their appearance. Although a number of Shang oracle bones mention eclipses, only one, on a turtle shell, can be seriously proposed as a total solar eclipse. It reads:

> Diviner Ko: ... day *i-mao* to [next] dawn, fog. Three flames ate the Sun. Big stars [seen].[24]

The date *i-mao* equates to day 52 of the sexagenary cycle and, in theory, we can identify any day of the cycle as far back as we can find evidence of its use. The Chinese specialists who translated the characters were able to find an eclipse on day 53 of a sexagenary cycle – one day out. This corresponds to *June 5 1302 BC*. This eclipse could have been total at Anyang, for an acceptable value of ΔT, but the identification is far from conclusive. Stephenson refused to use this eclipse in his study of ancient eclipses, finding no suitable Chinese records earlier than 720 BC.

If the Shang turtle shell were indeed a record of the eclipse on this date then, together with the secular change in the rate at about AD 700–1000, this might be our first indication that something had affected the diurnal rotation in an earlier era. If we had a whole series of eclipse records that were a day out then it might be possible to make such an assertion, but not based on a single disputed inscription.

The Kaliyuga Eclipse

In the Indian epic *The Mahabharata* there is another reference to a solar eclipse. It occurs on the eve of the great battle of Kurukshetra when the god Vishnu is said to have appeared before the combatants.

The Mahabharata, the "great battle of the descendants of king Bharata", is divided into eighteen books and is by some distance the longest poem ever written. Indian historians believe that it is, at least partly, historical and therefore many attempts have been made to date the battle by means of the eclipse. Whatever the historical reality, it is generally accepted that the literary arrangement of the epic continued to evolve up to about 100 BC. The Bharata battle brought the Rig Vedic age to an end and concludes the assimilation of the native Indian inhabitants by the Aryan invaders.

The Mahabharata contains some 150 astronomical references that are potentially of interest to a catastrophist researcher. Unfortunately some are contradictory and they cannot all be genuine. It is also well established that other Indian epics contain astronomical references relating to circumstances that must be astronomically dated hundreds of years apart.[25]

The conditions on the eve of battle are described by the sage Vyasa. Plants were blooming out of season and "sun moon and stars were all shining at the same time". We are also told that "Rahu seized the Moon" – a lunar eclipse.[26] The planet Saturn was near Rohini (the star Aldebaran in Taurus).[27] The narrator tells us that comets and meteors were seen. All the seven planets assembled in conjunction at the same place in the sky – the grand conjunction, which must itself imply a solar eclipse. Throughout the war, we are told that hundreds of meteors continued to pelt the earth.[28] At another time while two warriors are battling, we are told that Rahu eclipsed the sun and moon simultaneously.[29]

The astronomer Aryabhata in calculating the era of the Kaliyuga at 3102 BC (based on unknown astronomical sources) is generally held responsible for muddling the yuga system of the Rig Veda with that of the Puranas and the Mahabharata. This led to a widespread belief that the Bharata battle took place at 3102 BC.

So can the era of the Kaliyuga be astronomically dated by means of the eclipse reference? Unfortunately not; historians can count back the reigns and generations of kings and establish that the events in the Mahabharata could not be older than about 1400 BC and are probably even more recent. One estimate placed it at the eclipse of *1150 BC, 30 September*, based on the genealogies;[30] but more modern estimates suggest that this eclipse crossed Africa. Another possibility is the eclipse of *4 October 955 BC*; a small variation of ΔT would allow the path of totality to cross India.

The most likely candidate would be the eclipse of *4 July 857 BC*. Modern calculations suggest that the path crossed Iran, India and Thailand. The Sun and Moon were in Cancer with Jupiter close to the eclipsed Sun; and Saturn was in the horns of Taurus, near Aldebaran. However, there was no grand conjunction of the planets and the lunar eclipse on 18 July of the same eclipse season was not visible from India.[31]

We may see that some of the conditions are met and it is possible that an eclipse did occur around the time of a real battle, and was remembered. However, this has been augmented with so much older religious material taken from the Vedas, that we can no longer trust any of the astronomy.[32]

The Amarna 'Eclipses'

An extraordinary series of events occurred during the reign of the Egyptian king Akhenaten. In the first year of his reign, King Amenhotep IV abandoned the worship of the old Egyptian gods and announced that henceforth he would worship solely the Aten: a form of the sun-disk. He made Atenism the official religion and in the fourth year of his reign he renamed himself *Akhenaten*, which means "Spirit of the Aten". What is perhaps more surprising is that the people of Egypt went along with this heresy and indeed, it persisted into the reign of his successor Tutankhamun.

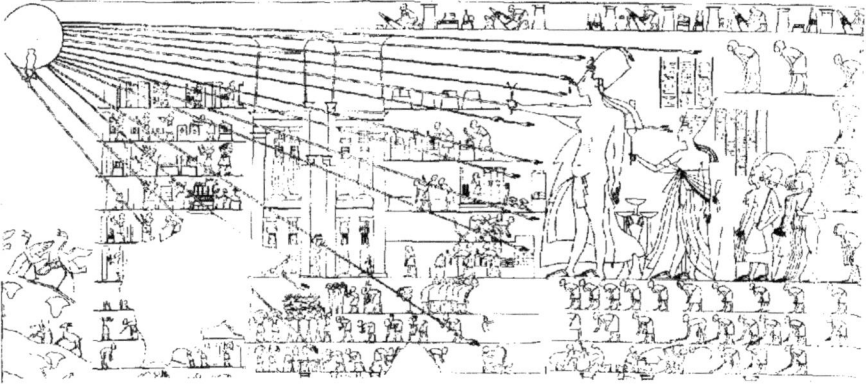

Figure 6.2 Akhenaten worships the Aten at Amarna.

Around the jubilee of his father Amenhotep III, the young Amenhotep IV was installed as co-regent in a ceremony that has come down to us in a temple inscription at Karnak, dedicated to the god *Re-Harakhti* – Re-in-the-Horizon, the rising sun. The inscription describes the new deity as "*Re Harakhti who rejoices on the horizon in his name of Solar light (Shu) which appears in the Solar Globe (Aten)*".[33] In a similar stele positioned between the paws of the Sphinx, Akhenaten's grandfather Thutmose IV claimed that in a dream, "Horus-in-the-Horizon" had spoken to him. That this was the Aten is confirmed by a commemorative scarab, stating that he had led his army into Palestine with the Aten before him. Thus, we see the origins of the later Atenist religion.

In the fifth year of his reign, Akhenaten as he was now called, founded a new capital city at Tel-el-Amarna on the eastern bank of the Nile. He named his new city *Akhetaten*, "horizon of the globe". The sole advantage of this particular patch of desert seems to have been that, at Amarna, the Sun could be seen to rise in a cleft in the mountains. An inscription on a boundary stele, declares that the site was chosen by the Aten himself, "in the place which the Aten enclosed on the eastern bank for His own self".[34]

Akhenaten's heresy was abandoned some eighteen years later by his successor, the boy-king Tutankhaten, who changed his name back to Tutankhamun and returned the royal court to Thebes. Akhenaten's statues and inscriptions were defaced and his name, and that of Tutankhamun, were struck from the king lists. It is probably this very fact that spared Tutankhamun's tomb from the looting that befell other royal tombs.

The dating of Akhenaten's reign, like much of the rest of Egyptian history, has to be reconstructed by counting the reigns from the nearest established Sothic date. The conventional chronology for this part of the Eighteenth Dynasty (ignoring co-regencies) is therefore as follows:

Tuthmose IV	1392-1382 BC
Amenhotep III	1382-1352 BC
Amenhotep IV-Akhenaten	1352-1336 BC
Tutankhamun	1336-1327 BC

However, a considerable margin of doubt attaches to the conventional chronology, particularly at this era; and some Egyptologists would like to revise these dates as much as three hundred years later. This is entirely a matter for Egyptologists to argue.

To return to the matter of eclipses, we may see that Egypt was visited by more than its fair share of eclipses during this dynasty.[35] The currently accepted values for ΔT show that a total eclipse crossed the Nile on *August 15 1352 BC* and would have been total at Amarna. According to the conventional chronology, this was the first year of Amenhotep IV/ Akhenaten.

However, on *May 14 1338 BC* another long eclipse (nearly seven minutes) crossed the Nile at Aswan and may just have been total at Thebes. It must have been spectacular. The conventional chronology puts this near the end of Akhenaten's reign. Only four years later, on *December 30 1332 BC* another eclipse was visible from Egypt. However,

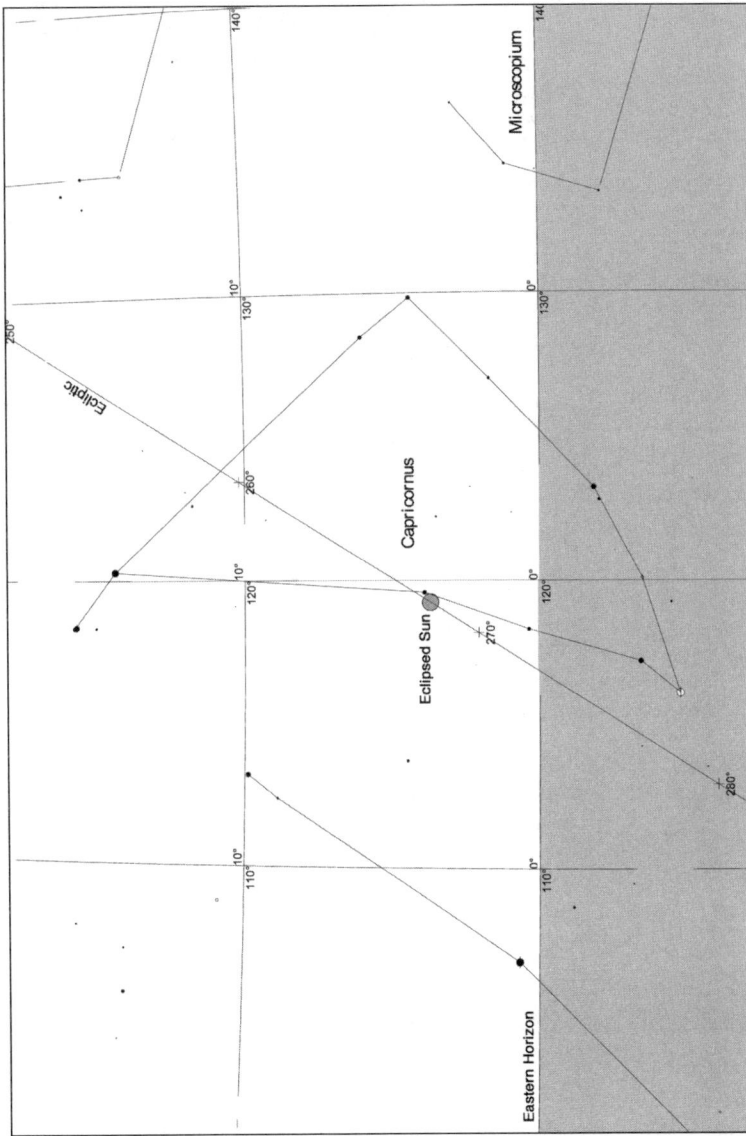

Figure 6.3 The Dawn Eclipse of 1332 BC

A dawn eclipse visible from the Nile valley between Cairo and Asyut on December 30 1332 BC. The sun rose in partial eclipse. For the eclipse to have been total at Amarna would require a slightly different value of delta-t from that currently accepted. (Retrocalculation using Skymap Pro 6).

this was a dawn eclipse. From a location midway between Amarna and Cairo, it would have risen only partially eclipsed and totality occurred just above the horizon. Again, on the conventional chronology, this was the fourth year of Tutankhamun's reign.

It is reasonable to expect then, during a period when the sun-disk was the prime deity, that these eclipses should have been liberally recorded in Egyptian inscriptions. This does not appear to be so. During the Amarna period, there are an abundance of reliefs and carvings from Akhetaten. These depict the sun-disk casting its rays down upon the king and his family as they worshipped the Aten. This raises the *reasonable suggestion*: could the Atenist heresy have been inspired by a series of eclipses?

One possibility is that the conventional chronology could be a few years in error and only a small difference in the value of ΔT would suffice to make the dawn eclipse total at Amarna. Could Akhenaten have been there to see it? Imagine the effect of such an eclipse upon this deeply religious man who already regarded the rising sun as his personal deity. Could it have been this sign, which compelled him to build his capital at Amarna?

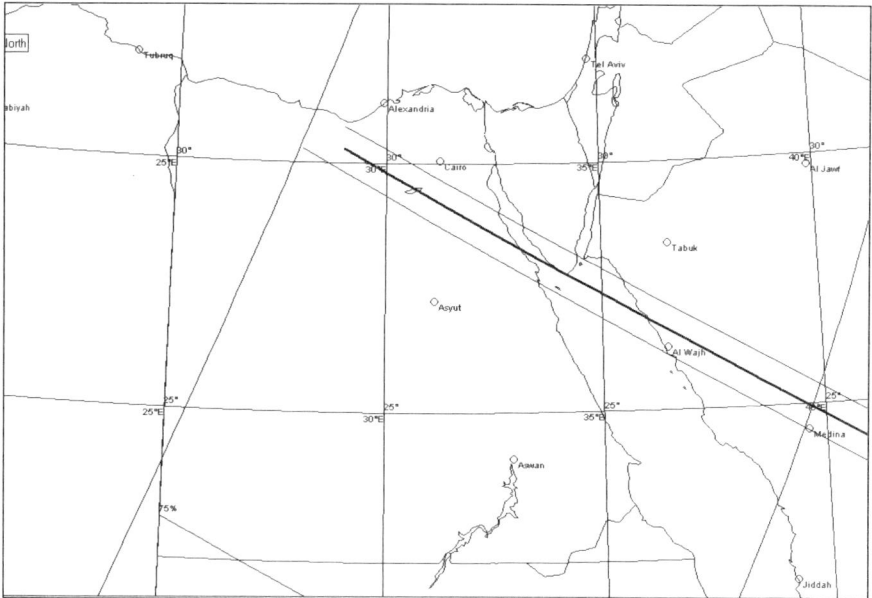

(a) 1332 BC

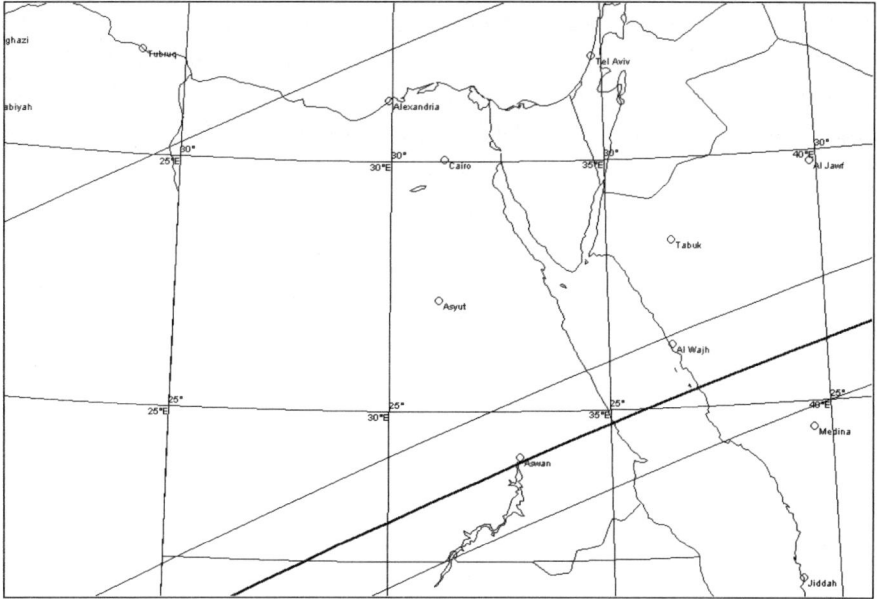

(b) 1338 BC

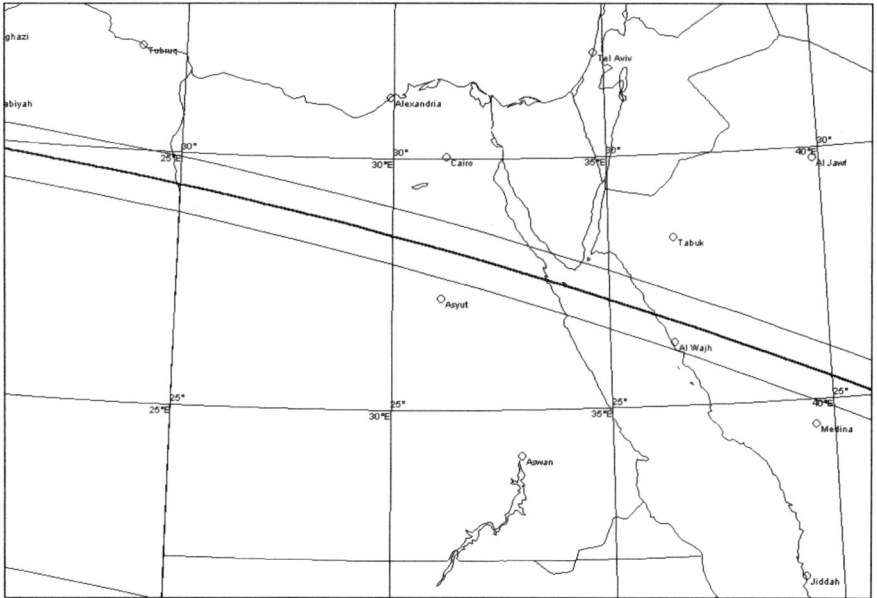

(c) 1352 BC

*Figure 6.4 The tracks of eclipses across Egypt during the Eighteenth Dynasty
(a) Eclipse 1332 BC, (b) Eclipse 1338 BC, (c) Eclipse 1352 BC.
(Retrocalculation using Skymap Pro 6).*

It is possible, as we have seen above, that if the dawn eclipse of 1332 BC fell in year 4 of Akhenaten's reign, then he could already have seen two previous total eclipses in his lifetime. Most of us are lucky to see one! This would at least explain why the Egyptian people went along with Akhenaten's heresy, as they would all have seen the same portents in the sky. If the Egyptologists could find an unambiguous inscription that mentions one of these eclipses then not only would it fix the Egyptian chronology, but it would also establish the stability of the Earth's rotation back to the fourteenth century BC.

A boundary stela from the city of Akhetaten contains a fragmentary proclamation, which Cyril Aldred rendered as follows:

> ...as Father Aten lived, something had been said which was more evil than that which the king had heard in his Year 4...more evil than what he had heard in his year 1...more evil than what King (Amenhotep III?) had heard...more evil than what king Tuthmosis IV had heard...[36]

What were these terrible omens and portents? And just what did Tuthmose IV hear?

The Venus Tablets

We have several times mentioned the Babylonian observations of Venus. Few historical documents can have caused so much controversy as these tablets. They were among the cache of cuneiform tablets discovered by Sir Henry Layard in 1850, during his excavation at the library of King Ashurbanipal in Nineveh (seventh century BC). When first translated in 1870, tablet 63 was shown to be a record of Venus observations; but it was not until 1875 that the year-formula of King Ammizaduga was recognised. They were copies of observations made during the twenty-one year reign of king Ammizaduga of the First Dynasty of Babylon.

Scholars quickly realised that this Venus data provided an opportunity to astronomically date the reign of king Ammizaduga. His dynastic ancestor, Hammurabi could be loosely dated between 1900 BC and 1650 BC by other synchronisms. Fixing Ammizaduga's reign would therefore provide a reference point by which the reign of Hammurabi and earlier reigns in the Sumerian king list could be determined. These lists take us back to Gilgamesh and the time of the Flood.[37]

The first and third sections of the Venus tablet hold observations of the planet's first appearance in the west and its last appearances in the east. For example, the first entry states:

> If on the 21st of Ab Venus disappeared in the east, remaining absent in the sky for two months and 11 days and in the month Arahsamna on the second day Venus was seen in the west, *there will be rains in the land; desolation will be wrought* [present author's italics].[38]

The middle section contains astrological portents for the periods of invisibility. For example:

> If on the 2nd of Nisan Venus appeared in the east, *distress will be in the land*. Until the 6th of Kislev she will stand in the east; on the 7th of Kislev she will disappear, and having remained absent 3 months in the sky, on the 8th of Adar Venus will shine forth in the west; *king will declare hostility against king*.[39]

The formulae are all somewhat repetitive and evidently, the copyist was more concerned with the recurrence of the astrological portents than with the observations themselves. It may be that such observations were routinely recorded in the reign of every king, but that those for Ammizaduga's reign survive solely because they were chosen as an example by an astrologer of the seventh century BC.

As the Venus tablets testify, the Babylonians were fatalists; all things came to pass solely by the will of the gods and humans possessed little free will. The intent of the gods might be foretold by astrological portents and it was for humans to understand them. Thus, an eclipse might foretell the death of a king, or alternatively the presence of Jupiter during an eclipse might imply that the omens were good.[40] History was of interest only so that these portents could be recognised should the circumstances repeat. Always there had to be an additional random element so that the diviner could never be proven wrong. For Mesopotamians, this would be an examination of sheep's entrails; for the Chinese, the cracks in a bone; or for the ancient Britons whether a hare, when released, runs to the left or to the right.

The complex rhythms of Venus have been discussed above in connection with the Mayan calendar, together with the relationship of the eight-year cycle of Venus to the eight-year lunar intercalation cycle.

It seems that the Old Babylonians were aware of the very same Venus cycles that the Maya recorded, but in this case, some two and a half millennia earlier. Yet, they never adopted the eight-year intercalation cycle as did the Greeks.

The cycle begins at superior conjunction when Venus lies on the far side of the Sun, at which time it is invisible for around three months. It then rises in the west, attaining its maximum eastern elongation in the evening sky. Maximum brightness is then attained before it fades again as it moves toward inferior conjunction with the Sun. After another, shorter, period of invisibility the morning cycle begins with inferior conjunction and the planet rises to its maximum morning brightness before attaining greatest western elongation. Finally, it fades as it moves back to superior conjunction. The entire cycle comprises the synodic period, varying between 577 and 592 days; average period 584 days.

Throughout the twentieth century, various scholars attempted to retro-calculate the dates when the risings of Venus would fall on the appropriate day and month of the Babylonian calendar. The theories and arguments can scarcely be done justice in such a short summary, but suffice to say that the scholars have not yet reached a firm conclusion. As the Venus-Moon cycle repeats, there are several sets of dates on which the observations could fall. These would place the first year of the Ammizaduga at 1702BC, 1646BC, 1638BC or 1582 BC, of which the currently most favoured date is 1702 BC.[41] Earlier scholars favoured even older dates, but these have been ruled out on archaeological grounds. Recently, the date 1419 BC has also been proposed. None of these options can give a precise match for the 29 and 30-day months, because the Babylonian method of intercalation followed no fixed rule.

Does anyone see what is wrong here? Scholars are attempting to use observations of Venus from the second millennium BC to determine an entire historical chronology for Mesopotamia and neighbouring regions. Yet, these same Babylonian astronomers, so the textbooks tell us, were incapable of devising an intercalation rule for their calendar and could not measure the true length of the year. How then, can we possibly rely upon their observations of Venus? It shows us again the selective blindness of archaeologists regarding the capabilities of ancient people and cultures.

The real question is surely *why* they were observing Venus so closely. If Babylonian astronomers of the second millennium could accurately record planetary movements then perhaps the Sumerian astronomers of a millennium earlier were equally capable. Could it be that their interest in Venus arose because they were using the planet as an absolute time standard and checking their irregular calendar against it? Why were they not content simply to count the calendar days between events?

For the Babylonians and Assyrians, perhaps more than for any other culture, eclipses were important portents; and eclipse prediction above all will not tolerate an inaccurate calendar. The astrologers had to be certain of astronomical time. An eclipse prediction that was a day out, might as well be a year out. It would be no use relying on priests to intercalate a month at the proper time; otherwise, they would run into the same difficulties that modern scholars have found. This would give us one reason why they might wish to tie their stellar calendar to a Venus rhythm. Another might be that, in the earliest period, the day itself was not reliable as a standard of time.

The king lists can be counted back from Ammizaduga to the kings who supposedly reigned at the time of the Great Flood and the mythical kings before it. Jacobsen, who collated the various fragments of the Sumerian king list, assumed an original list of which all the others were copies. At some point in this copying process, the list of 'antediluvian' rulers found in Berossus was appended. They are clearly of a different style. Some of the early kings are assigned incredibly long reigns and these have to be replaced with averaged reigns in order to make historical sense of the dynasties.

The astronomical dates for Ammizaduga enabled scholars to assert, with some confidence, that Hammurabi of the First Dynasty of Babylon ruled during the eighteenth century BC. Before the rise of Babylon, the dominant city in Mesopotamia was the Sumerian city of Ur. The demise of the third dynasty of Ur was foretold by a lunar eclipse; and some years earlier another lunar eclipse had been taken as an omen for the death of a king. Although lunar eclipses are generally less useful as dating indicators (as there are so many) this unique pairing enabled scholars to place these lunar eclipses at 2095 BC and 2053 BC. These are separated by forty-two years; and the Sumerian king list shows just such an interval between the death of a king named *Shulgi* and the fall of the dynasty.[42]

The various overlapping dynasties can then be counted back from these fixed points to give us a date around 2700-2900 BC for Gilgamesh of the Early Dynastic III period of Uruk. Still further back in time we find the references to the time of the Flood.

(In) Shuruppak Ubar-Tutu(k)
became king and reigned 18,600 years,
1 king
reigned its 18,600 years,
     5 cities were they
       8 kings
reigned their 241,200 years
The Flood swept thereover,
When kingship was lowered from heaven
The kingship was in Kish
In Kish Ga. .ur(?)
Became king
And reigned 1200 years;

Although the post-diluvian reigns are much shorter than the fabulous terms assigned to antediluvian rulers, even the 100 year reign claimed for Gilgamesh is still too long to accept. Replacing these with averaged reigns of thirty years per generation still leaves a considerable margin of uncertainty in these reconstructions – on top of that already conferred by the uncertainty of Ammizaduga's reign. Jacobsen derived a date of around 3100 BC for the dynasty of Kish and the Flood, based on this reconstruction. Modern scholars would probably wish to shorten this time scale by some 200 years and place a series of localised river-floods around the year 2900 BC.

Brothers Hsi and Brothers Ho

In the *Shu Ching*, we are told of the Chinese court astronomers Hsi and Ho during the reign of an emperor Chung K'ang (c2159-2147BC). These are apparently the descendants in office of those court astronomers, who were earlier charged to measure the year by the length of the shadow (see chapter 3 above). Sinologists are not sure, whether these are real astronomers named after the mythical brothers who supposedly turned back the sun at the points of solstice rising and setting, or whether this myth arose later from a memory of the real court astronomers.[43] At this early period, Shu Ching is almost legendary in character and the dates must be regarded as approximate.

The court astronomers were charged with predicting the astrological portents and initiating the proper ceremonies. We are told that: "the

former kings were carefully attentive to the warnings of heaven".
However, Hsi and Ho deserted their posts and indulged in drunken
pleasures. They failed to predict an eclipse:

> On the first day of the last month of autumn, the sun and
> moon did not meet harmoniously in Fang [*Scorpio*].[44]

An expedition was sent out to find and punish the aberrant
astronomers:

> The statutes of government say, "when they anticipate the time,
> let them be put to death without mercy; when their reckoning
> is behind the time, let them be put to death without mercy".

Now whether or not this document is an erroneous copying or a
later invention, it still tells us a lot. Firstly, as the source documents are
much older than the Han Dynasty, they clearly indicate that the Chinese
astronomers knew from an early period that eclipses were caused by the
meeting of Sun and Moon. It is also apparent that the astronomers were
predicting eclipses, not merely observing them, and were expected to
predict them with precision. If we take the reference at face value then
this knowledge dates to the third millennium BC. From the oracle bones
it is also clear that eclipses were being used to ensure the correct date of
the divination, not as portents in their own right.[45]

Next, we are told that the eclipse occurred on the first day of the last
month of autumn and that the Sun was in the constellation of Scorpio.
This would be true for a late-October or early-November eclipse around
2100 BC, but by the approximate date that Shu Ching was written,
precession had taken the late-autumn Sun into Libra. In order for later
chroniclers to invent this eclipse reference, they would have had to
understand the precession of the equinoxes. This again lends some
authenticity to the reported eclipse.

Various attempts were made to date the Hsi–Ho eclipse, ranging from 2165
BC to 1948 BC, before scholars gave-up on the reference as purely legendary.[46]
The dates most often cited are 2136 BC or 2137 BC. However, it should be
noted that nothing about the reference says that the eclipse was total.
Indeed the reference to the Sun and Moon "not meeting harmoniously"
would seem to imply only a partial eclipse. This gives a huge range of
possibilities, but *October 11 2136 BC* is the best possibility within a thirty-
year range of the traditional date. A reconstruction shows that it was just

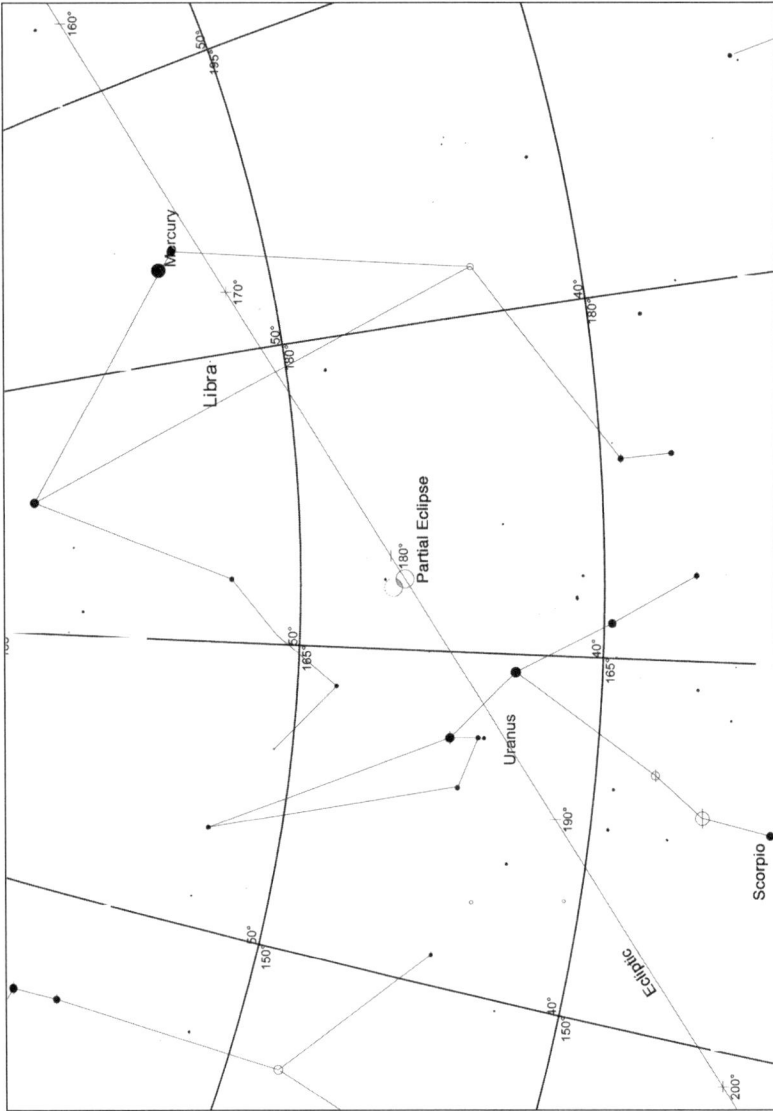

Figure 6.5 A New Moon in October
The circumstances of a partial eclipse as seen from northern China on 11 October 2136, BC. (Retrocalculation using Skymap Pro 6)

partial from about 6° north of Anyang, the later Shang capital. However, we can actually have no idea of the location where the eclipse was viewed.[47]

There are probably too many unknowns to learn very much from this eclipse. However, for the reasons stated above, we should probably regard it as a genuine report. It is again indicative of a high standard of astronomy in China during the third millennium BC; knowledge that appears to have been forgotten in later centuries. As we have seen, the Chinese legendary chronology is too uncertain for us to extrapolate back to the time of the Flood. However, it is safe to assume that the legendary Fu Hsi and Kung Kung must be placed several hundred years before the date of the Hsi-Ho eclipse.

Manetho's Eclipse

Attached to Manetho's Egyptian king list are a few comments about each king's reign. Manetho's sources are unknown; however, Egyptologists have noted that his style agrees closest with the oldest known Egyptian king list – the Palermo Stone, which dates from the Fifth Dynasty.

For Dynasty III, the first dynasty of the Old Kingdom, he supplies a very interesting comment:

> The Third Dynasty comprised nine kings from Memphis.
> 1. Necherophes, for 28 years In his reign the Libyans revolted against Egypt, and when the moon waxed beyond reckoning, they surrendered in terror.
> 2. Tosorthros for 29 years (in his reign lived Imuthes)…who was the inventor of building with hewn stone…[48]

The Armenian version of Eusebius gives this slightly differently and says that the Libyans were terrified when the Moon "waxed unseasonably".[49] What is this all about? Egyptologists usually treat it as a myth and ignore it; but this is a myth in a king list and king lists are the basis of history.

Unfortunately, nothing is known of the reign of this king Necherophes as this part is missing from the fragmentary Palermo Stone. Egyptologists therefore identify him with a king called Sanakhte, known only from archaeological finds; and he may be the same person as Nebka, who is mentioned in a story from Papyrus Westcar.[50] Egyptian kings bore many names and titles and their priorities changed around this time.

Egyptologists are in no such doubt about Tosorthros. He is identified with king *Djoser* who built the first pyramid, the step pyramid at Saquarra; and Imuthes is his famous architect *Imhotep*. King Djoser, or Zoser, is now usually regarded as the first king of the Third Dynasty. It may be that Sanakhte was another of those usurpers or heretics, who were omitted from later lists.

We therefore have a reference to some abnormal variation of the Moon right at the beginning of the pyramid age. One has to trust the translators, for the text has undoubtedly passed through more than one translation. We would use "waxing" to imply the rising or brightening of the Moon during its monthly phases, the opposite of "waning", the darkening or setting.

The reference could be to a lunar eclipse, but this implies a darkening, rather than a brightening of the Moon and lunar eclipses have seldom been a cause for general alarm. The term "waxing unseasonably" could imply a failure of the calendar, whereby festivals and portents were drifting out of their proper season. This is certainly around the era at which Egyptologists place the foundation of the civil calendar, but it seems unlikely that soldiers would surrender in terror just because of a faulty calendar.

It is reasonable to suggest that this is yet another reference to a solar eclipse during or on the eve of a battle. The fact that it was taken as a bad portent by the Libyans would further suggest that the eclipse was total along the north coast of Africa, but only partial further south on the Nile. As the reference is in a king list we should be able to find it and date it in the same way that dates are assigned to the reign of a king.

The uncertainty of the Egyptian chronology at this era widens to about 100-150 years; and similar uncertainty in the calculation of ΔT could place the calculated eclipse paths on the opposite side of the planet. An astronomical date this early would fix the chronology for the pyramid age and would take us back with a little more certainty towards a historical date for the Flood.

The conventional dates for the Third Dynasty place it at 2686–2613 BC so we would have to seek the eclipse somewhere in the range: 2686BC±150 years! One outside candidate might be the eclipse of *9 October 2760 BC*, which crossed Cyrenaica and Sudan.

Another opportunity is afforded by the astronomical alignments that have been claimed for the so-called 'airshafts' of the Great Pyramid of

Giza. For example, it has long been known that the sloping entrance passage of the Great Pyramid of Khufu (Fourth Dynasty) was oriented towards the celestial pole. During the pyramid age, the star α-Draconis would have been close to the celestial pole. Author's Bauval and Gilbert have also suggested that the pyramid's southern 'airshafts' were aligned towards the culmination of Sirius and the three stars in the belt of Orion. From these alignments they would date the building of the pyramid to approximately 2450 BC.[51]

A study by Egyptologist Katherine Spence took a more extensive look at the alignment of all the Old Kingdom pyramids and monuments. She demonstrates that the approximate north-south alignment of these monuments was achieved using a chord of simultaneous transit drawn between two circumpolar stars.[52] Her conclusion is that the most likely date for this chord to have crossed the true pole, based on the rate of precession, was 2465 BC. As the archaeological dating of Khufu's reign places him at 2540 BC, this astronomical evidence would suggest that the conventional chronology for the Old Kingdom should be lowered by some seventy-five years.

A revision based on this astronomical evidence would therefore place the start of the Third Dynasty around 2610 BC. Therefore, a 150-year range either side of this date further extends the already wide range in which we would have to seek Manetho's 'eclipse'. The best candidate within this revised range therefore becomes the eclipse of *August 10 2654 BC*. This eclipse had a west to east track that runs all along the North African coast and, assuming the standard value for ΔT, would fulfil all the conditions for Manetho's eclipse.[53] This is again a remarkable coincidence.

However, Manetho's reference does not state in which year of the king's reign the Libyan revolt occurred. All we know is that it could have occurred anytime within the twenty-eight year reign of a little-known Pharaoh, Sanakhte-Nebka. It still leaves us with a wide twenty-eight year zone for the start of the Third Dynasty and the Old Kingdom, which would place it between 2672 BC and 2644 BC.

The beginnings of Egyptian history can be estimated by counting back however many years we wish to assume for the First and Second Dynasties. The conventional chronology allows 204 years for the Second Dynasty and 210 for the First, which takes us back to 3100 BC and the reign of Menes, the unifier of Upper and Lower Egypt. The present

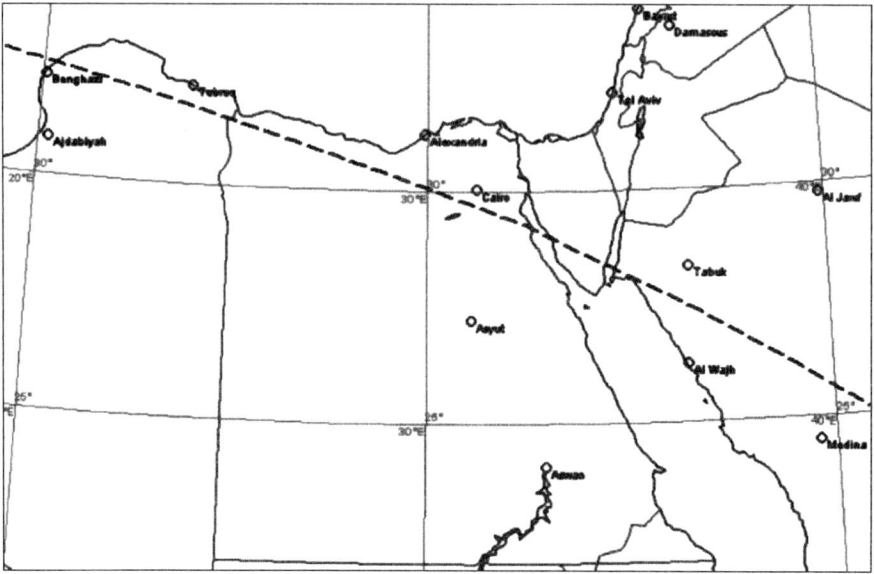

Figure 6.6 Manetho's Eclipse
The approximate track of an eclipse that occurred on 10 August 2654 BC, which passed all along the Libyan coast and across the Nile Delta. According to Manetho, "the Libyans surrendered in terror when the Moon waxed beyond reckoning" (reconstruction based on Redshift 2).

author's suggested identification of Manetho's 'eclipse' would do little more than offer confirmation of the chronology that most Egyptologists accept, giving a range between 3086 BC and 3058 BC.

Before Menes, we have only a few names from the Palermo Stone for kings of the predynastic era, before Egypt was unified, or alternatively we have the gods and demigods of Manetho's list. Manetho, or rather his Jewish and Christian commentators, placed the Biblical Flood immediately before the reign of Menes, although there is little in the details of his reign to indicate such a calamity. We should not rule out a dark age of perhaps a century or more before the unification. The only certainty seems to be that by the Fifth Dynasty, when the Palermo Stone was written, the Egyptians themselves remembered little more about this period than we know today.

So it may be seen that, the further back in time we go, the more difficult it becomes to prove both historical chronology and the stability of the Earth's rotation by astronomical methods.

Eclipses and Catastrophism

It may be appropriate to restate at this point, *why* eclipses and other astronomical dating methods are of as much value to students of catastrophism as they are to archaeologists and historians.

The uniformitarian position assumes that the conditions on the Earth today are more or less as they have always been and that the only forces that apply are those which apply today. Therefore, once we have established a calendar: the day, the month, the year; then we can simply extrapolate these as far into the past as we wish to go. The tidal interaction of the Earth–Moon system introduces a minor variation, but this too is uniformitarian and can be estimated by rules. Catastrophism however, holds that present conditions would only apply as far back as the most recent catastrophic episode; before this time we can no longer work out what the world was like simply by projecting back today's conditions.

One principle that is not always appreciated by catastrophists of various kinds is that any cataclysm that could cause the Great Flood *cannot be localised*. It would affect the entire world and its effects must be simultaneous world-wide.

The Earth's rotation is today in a state of balance. It rotates about its axis of figure: the north-south axis. If the waters of the Mediterranean suddenly burst into the basin of the Black Sea, or if a lost continent collapses into the depths of the Atlantic Ocean, then this *must* disturb the rotational balance and the world would wobble on its axis. Similarly a comet or asteroid impact in the ocean must cause the Earth to wobble; producing secondary effects upon weather and sea level that would be evident world-wide. It is these secondary effects that are remembered all over the world in the various flood myths.

The popular theory of crustal slippage is often advanced as an explanation of pole-shift and catastrophist phenomena. However, the cause is never explained. Even Hapgood, who proposed it, ducked the question of a mechanism, citing only vague 'imbalances' in the crust.[54] What does imbalance mean? It means that the Earth is not symmetrical or is not rotating about its figure axis. In other words, it must wobble!

Geophysicists however, do understand how our planet behaves when its rotation is slightly disturbed by internal or external forces. However, they are uniformitarians and few numerate scientists would admit myths as evidence. They seldom equate the theoretical constructs in their equations with events on the real Earth. They know that if the rotation is disturbed then it must wobble in two theoretical modes.

The first mode is the short-lived Chandler wobble, in which the axis wanders within the body of the Earth as it hunts for the stable position. This motion, if it were of a significant amplitude, would cause pole tides and all the other effects that we find associated with a world-wide flood. However, all this would be completed within about twenty years.

The second mode is the core mode, in which the rotation of the outer-earth becomes misaligned from that of the core of the planet. In this case, the motion is mainly in space and it should not cause pole tides and earthquakes. However, it would persist for a very long time, perhaps for two or three millennia. Geophysicists are unsure of the lifetime of this, largely theoretical, motion. Its effects would be mainly on the long-term climate (as the seasonal height of the Sun would vary) and it would affect all astronomical observations, including the *latitude* of ancient eclipse sightings.

Even if the length of the day is not directly changed in the initial catastrophe then it must be affected indirectly. It would therefore be pointless to project back the value of ΔT into a period when the Earth was wobbling. The identification of eclipse paths back as far as 700 BC tells us that the length-of-day was stable this far back; but if some other non-tidal influence was present before AD 700-1000 then this could indicate recovery from an ancient transient event. An exponential rate of recovery would mean that most of the damaging effects would manifest in the early years after the initial dislocation, tailing off gradually.

If we could identify eclipse paths and determine ΔT further back into history then it would push back even further the date when a catastrophe could have occurred. Furthermore, if we could determine an ancient eclipse path that was even a fraction of a degree north or south of its calculated track then this would be evidence of a wobble or a pole shift. Tidal slowing alone could only affect the longitude.

Another interesting possibility is that a change to the length of day, or a pole shift, could also affect the rate of precession of the equinoxes. Precession is directly proportional to the cosine of obliquity. Therefore, when we find ancient alignments that can be dated by the precession, then it is interesting that they should give a different date to that supplied by purely archaeological methods. It may imply that the archaeological date is correct, but that the obliquity has changed (or was in the process of changing) since the pyramid age.

That which the generalist can suggest is for the specialist to prove or disprove. What the specialist may not do however, is to dismiss and ignore. This however, is just what academic specialists have been doing to mythological evidence for as long as specialists have existed. Historical sources and astronomical retro-calculation methods are currently too uncertain to prove an ancient catastrophe, but for the same reasons, they are equally unable to disprove it.

References to Chapter Six

1   Stephenson, F.R., 1997, p7

2   Stephenson and Holden 1986, introduction, iix

3   The perturbation occurs because the Sun's gravity is stronger when the Moon is closest to the Sun at new moon and weakest at full moon.

4   Meeus, J. 1982, J.Brit. Astr. Assn., 92,124-6

5   For a discussion of other eclipse cycles, see Aveni, A., 1997, p 35-6

6   The earliest application of the name *Saros* for the eclipse cycle was by the Tenth century Greek lexicographer Suidas.

7   Aetius, 2,24,3

8   Plutarch, Life of Pericles, 35; translation by I. Scott-Kilvert

9   Ronan & Needham, 1981, p203

10 As the Assyrians did not use an era as such, each year was named after a minor religious official, who kept a record of the notable events during his year of office.

11 Stephenson, F.R. 1997

12 For example, all the retro calculations performed for this study were prepared using *Skymap Pro 6*, which utilises the formulae for delta-T given in the paper by Stephenson and Houlden (1986) for which the accuracy is estimated as 15 minutes at 1500 BC. Wherever possible these have also been checked using *Redshift 2*.

13 Lambeck, K, 1988, p 607 & 636-7

14 Stephenson, 1997, p 517

15 Herodotus I, 74

16 Pliny, Natural History, II,53

17 Thucydides, History of the Pelopponese War, II, 28

18 Stephenson, 1997, p346-348

19 de Jong, T & van Soldt, W.H. 1989 p 239

20 Pardee, D. and Swerdlow, N., 1993 p 406

21 See Mitchell, W.A. (1990); also Sawyer and Stephenson (1970)

22 Sawyer and Stephenson (1970)

23 Rohl, D, 1995, pp 237-241

24 Stephenson, F.R, 1997, p218

25 See Koshar, R,. (2000) pp 25-26 regarding references in the Satapatha Brahmana dating from 2300 BC and 1400 BC.

26 Rahu, or Asura, is the Hindu demon who devours the Sun or Moon during an eclipse.

27 Bhisma parvan 3,10; 2,32; 3,11

28 Bhisma parvan 17,2; 19,38

29 Asvamedha parvan 76,15

30 Archdeacon Pratt's method – see Sita Nath Pradham, Chronology of Ancient India, 1996, pp 268-271.

31 Calculations using Skymap Pro 6.

32 See for example, the story of Atri and the Lost Sun; O'Flaherty, 1981, p187

33 This translation is given by Desroches-Noblecourt in *Tutankhamun*, p 92.

34 Aldred, C., 1988, p 49

35 Retro-calculations based on Skymap Pro 6

36 Aldred, C., 1988, p 50

37 The reconstructed king-list is given by Jacobsen, 1939

38 Tablet K.160, obverse; translation by S. Langdon, 1928, p 7

39 Ibid, p 10

40 See Grayson , A.K. and Redford, D.B., 1973, pp 81-82

41 P.J. Huber, 1987, pp 5-17.

42 The more precise dates are: January 20 or July 25 2095 BC; and April 13 (partial) or October 8 2053 BC.

43 Ronan (Needham) 1981, p 75

44 Translation by James Legge, See Waltham, C, pp 66-61

45 Ronan (Needham) 1981, p197

46 Ibid, pp194-195

47 Retro-calculation by Skymap Pro 6

48 Manetho, fragment 11, from Syncellus via Eusebius; translation by W.G. Waddell.

49 Manetho, fragment 12, from the Armenian version of Eusebius.

50 Regarding this Nebka, see Simpson, W.M. 1973, p 16-19 where he is referred to as "Nebka the vindicated".

51 Bauval, R and Gilbert, A, 1994, pp 179-182

52 Catherine Emma Spence, unpublished thesis 1997. True north can be established by hanging a plumb line to mark the meridian and noting the direction at which both stars simultaneously align with the chord.

53 This eclipse can easily be reproduced using Redshift 2 software. Skymap Pro gives a slightly different path.

54 Hapgood, C, 1958 (republished 1999) pp 40-44

7

The Problem of Time

Consider that the end of the world arrives suddenly one day and you are fortunate enough to be among the few survivors. Would you remember the date? You might remember the day of the week and the month of the year. Perhaps someone in your little band would recall the day of the month and you begin, like the stranded Robinson Crusoe, to mark-off the passage of the days. With all the priorities of survival and amid unseasonable storms and extremes of weather, years might pass before you realised that the Sun and the Moon did not conform to the expected seasons; and that the world had changed. Your calendar no longer works. How would you set about working out what had gone wrong with the world?

Something like this, must have befallen the survivors of the cataclysm we call the Great Flood. We need not take the myths too precisely. It is clear that whatever the nature of the catastrophe, many ordinary people survived, in many places, and they tried to carry on as best they could. We have seen that the earliest calendar traditions recall a year of 360 days. Even a lunar calendar would soon begin to drift if the length of day changed and the aberrant Moon would surely be noticed even before the gradual drift of the seasons.

The Flood and the Earth's Wobble

The Flood, *as it is described* in the various religious texts and myths *must* be associated with a nutation of the Earth's axis of rotation. If the ocean spills over the land (for whatever reason) then this would alter the symmetry of the planet and must reposition the axis of figure within the Earth. The rotation then becomes unbalanced and it would wobble. Conversely, if the Earth were set to wobble on its axis due to some external influence then this in turn must cause the sea level to fluctuate and transgress the shoreline. This is the phenomenon that geophysicists call a *pole tide*.[1] It was one of the first scientific explanations advanced by nineteenth century scholars to explain the Biblical Flood.[2]

The sea level world-wide is determined by the balance of gravity pulling the sea into its spheroidal figure; and the centrifugal effect of the diurnal rotation, which tends to throw it outwards at the equator. Thus, the shape of the Earth is an irregular flattened ellipsoid, known as the *geoid*, with an equatorial radius that is some 21.4 km (or about thirteen miles) greater than its polar radius. The Earth therefore has an axis of symmetry, termed the *axis of figure*, which coincides closely to the present-day north-south axis. Stable rotation occurs only when the planet rotates about its figure axis, a condition that very nearly applies to the Earth today.

However, this rotational balance is conditional upon the symmetry of the geoid. Earthquakes, or ocean water sloshing around on the surface, would alter the figure of the planet and drag the figure axis away from the axis of rotation. The height of the land on which you stand is determined by the axis of figure, while the sea level must remain nearly symmetrical about the instantaneous rotation axis. This dynamic relationship between the shape of the solid earth and the ellipsoid of the oceans is geologically ancient. If the rotation should become unbalanced then, at any given location, the sea may either recede or spill over the shoreline; rivers would flow erratically and lakes would slosh around in their beds.

What would a pole tide be like for a person standing on the seashore? One is tempted to visualise it like a devastating mega-tsunami: the sea recedes from the shore before returning in a great wall of water – but it would not be like that at all! Rather, the sea would just become tilted by a fraction of a degree all over the world. Within the period of a day it would flow downhill to find the lower level. The tide would appear to be gently flowing in as it always does – but then it would just keep on coming, until you were completely submerged. Equally, it might ebb away from you, leaving the sea bed temporarily exposed. Over a roughly 14-month period it would then range between these two extremes.

Remember too, that the pole tide also affects the solid earth. As the continents tilt and the land rises or sinks, then the seashore would eventually return very close to its former shoreline. It would leave little spectacular evidence, other than a pattern of raised beaches and drowned forests around the modern shoreline.

Thus, it may be seen that the Great Flood as described, if it is not caused by a nutation of the axis, must itself cause such an excitation. This conclusion is inescapable. Therefore, if the Flood were a real event then it should be possible to find some evidence of this axial wobble in the physical record of the past.

All of this is standard geophysics, although the reader will seldom find it so plainly summarised. The simplest mode of rotational oscillation is that discovered by Chandler in 1892. It has a period of approximately fourteen months and combines with the seasons to give a roughly seven-year rhythm. Geophysicists calculate that the Chandler wobble should decay exponentially and, unless something re-excites it, the motion should be fully damped by the viscosity of the Earth's interior within a period of 20-40 years. Although the pole tide associated with the Chandler wobble is scarcely detectable, its period is completely independent of the amplitude.

Exponential decay is a very powerful effect. A lifetime of twenty years implies a rapid falling away of the effects followed by a gradual tail-off. The worst effects of erratic sea levels would pass within the first year and the Biblical 'forty days', or even the seven days of Gilgamesh are not entirely to be ruled out. Even a wobble of a fraction of a degree of latitude could potentially generate a pole tide that would submerge most low-lying regions of the world.

The Flood must also be associated with a permanent pole shift – but this need only be of the order of a fraction of a degree of latitude to produce the effects described. If the Earth deforms and the figure axis migrates then the poles would move on the Earth's surface. Balance can only be restored when the world again rotates about the new axis of figure. So again, if the wobble were not caused by a pole shift then the rotational wobble must itself trigger such a shift. Even the tiny modern Chandler wobble causes such a small but measurable permanent shift.

However, this leaves open the question of what could cause the initial displacement of mass that disturbs the rotational stability. The Chandler wobble does not affect the attitude of the Earth's principal axis in space, the *obliquity*, as the angular momentum remains unchanged throughout.

There is a second mode in which the Earth could wobble; and this arises because our planet has a fluid nickel-iron core. Rapid changes, such as those considered above, cannot be transmitted through a fluid and the core therefore carries on rotating as before. Today, the principal rotational axes of the crust and mantle, and that of the core, are very closely aligned. Should the inner and outer rotations become misaligned then the two axes must spiral about each other. This 'core mode' is vanishingly small on the Earth today.

The core mode is subject to only very weak damping, due to the lubricating effect of the fluid core. Once triggered, it should persist for a long time, possibly for thousands of years. As the present-day rotation is stable, there is little data to go on and we must rely upon the theoretical constructs of the geophysicists. However, if the Earth possesses one mode of oscillation then the equations tell us that it must also possess the other.

Nothing rules out that the Chandler wobble of the Earth could be excited again, while the core mode was still recovering from a previous episode. Indeed this is quite likely, given that the interior would remain in an unstable state.

So again, if the Great Flood were a real event then it should also have triggered the core mode. As this would persist through thousands of years, we should expect to find some evidence of this too in the ancient data. Moreover, since it is primarily a motion of the axis *in space* it should reveal itself in ancient astronomical observations (such as eclipse records and alignments) and in the climate record.

What then of the length of day? In addition to the tidal effects of the Moon, geophysicists detect unexplained variations of the order of milliseconds known as *decade fluctuations*. However, the considerable changes to the length of day that are implied by the ancient calendars could only be caused by a significant external force exerted upon the Earth. For the moment, we must set aside the cause of such a change, as most scientific opinion would probably consider it impossible.

However, we may safely assume that an external force on the surface would affect only the outer shell (the crust and mantle) and a period of equalisation of the inner and outer lengths of day would follow. Would this be a short period of twenty years like the Chandler wobble? Alternatively would it be a slow process, like the core wobble? Only the specialists could answer such a question and as it is usually considered impossible, no opinion can be cited. However, the stability of the period of the day back as far as Old Kingdom Egypt, as considered in the previous chapter, would suggest that any such transition must take place rapidly.

Megaliths and Astronomy

The evidence discussed in previous chapters has pointed towards a conclusion that a major cataclysm occurred in the late fourth millennium BC. The tendency for oral traditions of similar events to degrade into a single myth means that the memory of this event may incorporate aspects of more recent events and characters, as well as distant memories of older episodes. There has probably *always* been a Flood myth.

The suggested date of c.3100 BC for the Flood comes from the Indian and Meso-American calendar eras, but really it only echoes the conclusions of the early Christian chronologists, suitably confirmed by modern science and other evidence. The chronologies of China and Mesopotamia also indicate a cataclysm around this date; and the more verifiable chronology of Egypt suggests that it could not be more recent than 3050 BC. For most other areas of the world we possess only timeless myths, but for western Europe at least, we also have monuments. Moreover, some of these monuments may preserve ancient astronomical alignments. Might these offer further clues to the past?

The oldest stone structures built in western Europe are those found in Iberia and Malta, which can be dated to as early as 5500 BC. Further north, in the British Isles and western France, they first appear around 4700 BC, along with the earliest evidence of a farming culture.

The Early and Middle Neolithic chronologies for Britain and Ireland had to be redefined following the advent of calibrated radiocarbon dates. The Neolithic transition is now placed at c.4700 BC and reached a peak of activity between 3600 and 3200 BC.[3] The monuments from this period cluster around the western coasts, where megalithic tombs with a distinctive crescent-shaped forecourt are encountered.[4] Aubrey Burl, who has extensively studied British Neolithic monuments and the related *allee-couverte* tombs of Brittany, considered that their passages were only loosely aligned toward sunrise or moonrise, perhaps at the season of their construction.[5]

In the late fourth millennium BC, a population decline appears to have set in, which some archaeologists have termed the "Mid-Neolithic Crisis".[6] Dateable artefacts from the period 3200 BC to 2900 BC are relatively rare.[7] The long barrows and horned-cairns of the Middle Neolithic ceased to be built. Throughout Ireland, many early field systems were abandoned, soon to disappear beneath a blanket of peat. There are even indicators of a population replacement around this time.[8] The slender long-headed race of the Middle Neolithic was succeeded by the stocky round-headed population of the Late Neolithic and Bronze Age.

The oldest evidence for an astronomical alignment was discovered during the excavation of the Newgrange mound in Ireland, the largest of the complex of mounds constructed on the bend in the River Boyne. The entrance passage leads to a cross-shaped chamber at the core of the mound. Along the chamber walls, the lining slabs are decorated with a variety of spiral and zigzag motifs; and an old legend claimed that, at a certain time of year, a sunbeam fell upon a three-leaved spiral carved on the back wall of the chamber. The Newgrange mound was radiocarbon dated from charcoal found between the excavated stones, which gave a date around 3150±100 BC for its construction.[9]

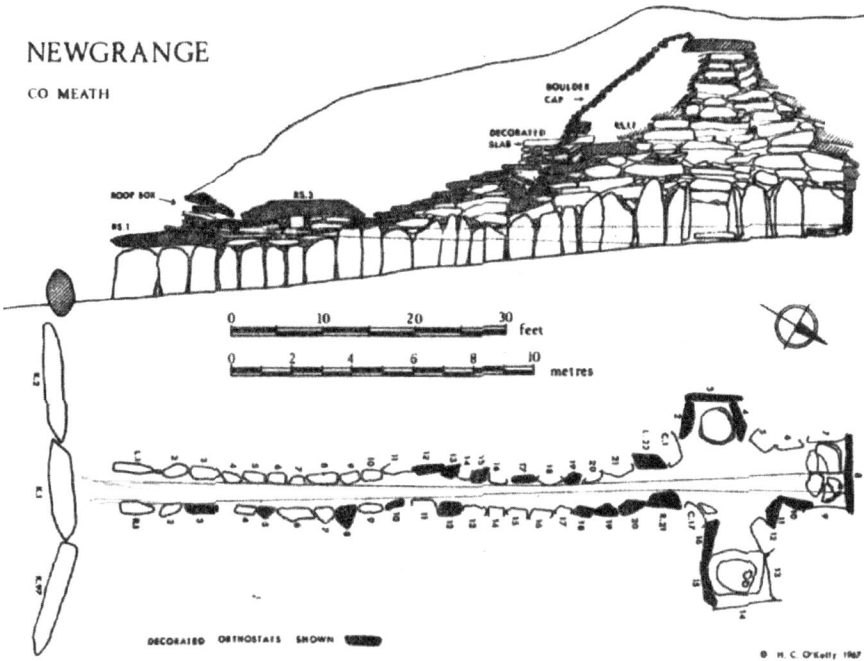

Figure 7.1 Winter solstice alignments at Newgrange
108 Newgrange, Ireland, after J.D. Patrick and H.C. O'Kelly (Patrick 1974, 517). The roof-box is a gap above the lintel stone at the threshold. At midwinter, as the sun rises, the sun illuminates the chamber by the beam shown shaded in the section (top) and plan.
Key: RS, roof slabs; K. kerb stones.
Reproduced from J. Patrick, Nature, 1974

The passage at Newgrange, and those at similar monuments such as Bryn Celli Dhu on Anglesey, are oriented towards the winter solstice sunrise. This alignment was first noted by the astronomer Sir Norman Lockyer.[10] However, during excavation at Newgrange during the 1960's, a slit known as the "roof-box" was discovered above the entrance. Shining through this slit, the rising sun of five thousand years ago would have illuminated the rear wall of the inner chamber during the days of winter solstice.

Arguments that the Newgrange alignment is mere chance stem from the fact that the roof-box is so wide that the sun would still illuminate the chamber if its declination fell within a range of three degrees of its calculated position. Rather than being an argument against the alignment, this is just the aperture that would have been required to track the Sun if the rotation axis were wobbling. The sun would then have risen slightly to the north or south of the calculated mean position, sending a narrow beam along the walls of the chamber. The range of possible azimuths (between -22°58′ and -25°53′) set a limit to how large such a motion could have been.

J. Patrick, who surveyed the Newgrange alignment, concluded:

> It therefore seems that the Sun has shone down the passage of the chamber ever since the date of its construction and will probably continue to do so forever, regardless of secular changes in the obliquity of the ecliptic.[11]

Secular variations of the ecliptic are one of the long-term factors that cause the Earth's ice ages. The present author is here considering a free wobble of the axis, of much shorter duration, but a similar principle applies.

A more recent survey of the winter solstice phenomenon concluded that the alignment is even more precise and was almost certainly the designed purpose of the monument. Newgrange is therefore the oldest megalithic monument for which an astronomical alignment can positively be claimed.[12]

In southern Britain around the same time, the people began to build near-circular enclosures surrounded by a deep ditch and a bank, known as *henges*. They were not obviously defensive in nature, but the ditch must surely have been intended to act as a barrier, perhaps to keep out animals or robbers.

Some henges may have contained rings of posts, or circular timber buildings, but others never contained any obvious structure at all. The most celebrated examples of this architecture are Durrington Walls near Stonehenge and nearby Woodhenge. Both once contained timber buildings. A henge at Arminghall, Norfolk may have held a horseshoe-shaped structure of oak-tree trunks; similar to the ring of fifty-five upturned tree-stumps preserved at the so-called Sea-henge site at Holme-next-the-Sea, on the Norfolk coast.[13]

From the late fourth millennium BC, through to around 1500 BC, stone circles were constructed throughout the British Isles, with a few more in western Brittany. The precise purpose of these monuments remains an enigma. It is interesting that the most celebrated of them, the rings at Avebury and Stonehenge, Wiltshire, were placed within circular

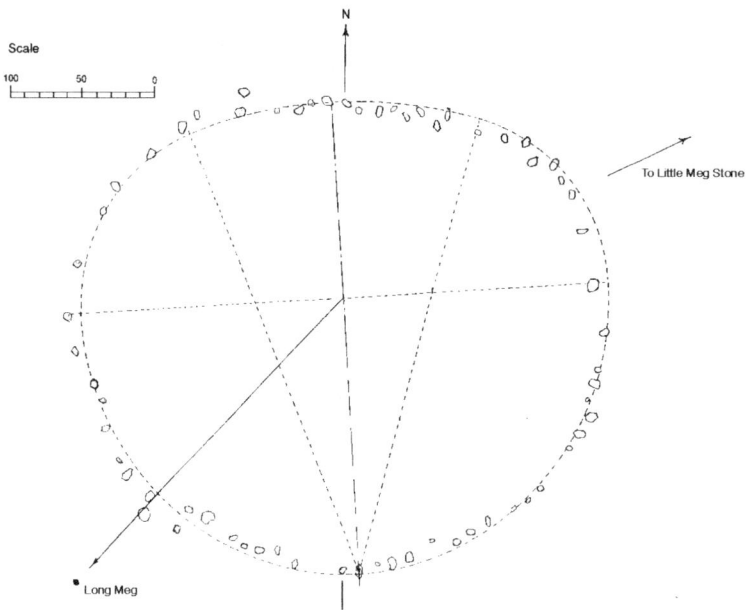

Figure 7.2
A plan of the stone circle known as Long Meg and her Daughters, Cumbria, England. The circle seems to be aligned on an axis a few degrees to the west of true north. Could this perhaps indicate that the poles have shifted since it was built? It is also interesting that the Long Meg stone carries a spiral carving, which may represent the motion of the celestial pole as the Earth wobbled on its axis.

henges. This would argue that the two types of monument served a similar ritual function, regardless of whether the circle was built of stone, wooden posts or upturned tree stumps.

The oldest stone circles, date to around 3000 BC and occur in highland and coastal areas, such as Castle Rigg, and Long Meg in Cumbria; and at Stenness, Orkney. These may predate by a generation the southern circles at Avebury and the earliest phase of Stonehenge.

The most unique of the stone circles is that at Callanish on the Hebridean Isle of Lewis, for in addition to the circle itself are four stone rows that run out towards the cardinal points. The longest avenue is a double row of stones pointing to the west of north. Looking south, the western row of this avenue is aligned towards Mount Clisham, the highest point on the island. This also corresponds to the point of maximum midsummer moonset for about 1500 BC. Another alignment points to the rising of the Moon, while the southern row gives the direction of maximum altitude.

Captain Boyle Somerville, in 1912, noted that the east-west alignments intersected with the centre line of the northern avenue; and thought that this was aligned towards the rising of the star Capella.[14] He also remarked that the eastern row was aligned to the rising of the Pleiades – a Neolithic equinox marker; and he suggested that Callanish rather than Stonehenge, might be the "spherical temple" described by Hecataeus. Hawkins however, viewed these alignments as yet further support for his eclipse prediction theory.[15]

Just as there is an Arctic Circle for the sun, so there is similar limit for the Moon, but due to the 18.61-year cycle of the Moon's precession, this oscillates between a northern and a southern extreme. This was discussed in Chapter 5, in connection with the alignments at Stonehenge. The points of the Moon's rising and setting on the horizon therefore swing between the points of major and minor standstill. As Callanish is further north than Stonehenge, it comes very close to the Arctic circle for the Moon. Every 18-19 years at its southern extreme, the midsummer Moon will be seen to rise only about one degree, just skimming the southern horizon. There is indeed remarkable correspondence with Hecataeus, who says that, at the spherical temple of the Hyperboreans, the Moon came very close to the Earth once every nineteen years.

Modern investigations at Callanish assign a date of around 2200 BC for its construction, but the outlying avenues may be later additions.[16] An alignment of the east-north-east avenue towards the Pleiades demands a date of around 1550 BC for the addition of the avenue – if the Pleiades are indeed the target. About a thousand years later the site began to be enveloped in a layer of peat, its purpose evidently forgotten.

Pleiades alignments

Other monuments too, may exhibit stellar alignments. As with the Egyptian use of the star Sirius, the heliacal rising of any star or constellation can be employed to mark the season of the year. This may not seem very important to modern people with their calendars and wristwatches, but for ancient people operating a lunar calendar, then the calendar date of a solar festival would vary from year to year. Ordinary people, or even priests, might not agree on the proper day to hold their festivals.[17]

It was Lockyer again who highlighted many of these alignments. In many ways, his work is archaeoastronomy in its pure form, untainted by 1960's theories of eclipse computers. His contemporaries were severely critical of his ideas because he seemed to want too many alignments. Archaeologists almost universally suppressed his astronomical theories until they came to the fore again after of the work of Hawkins and Thom.

Several ancient stone rows survive on the moors of Devon and Cornwall. Two such avenues stand at Merrivale on Dartmoor. The northern avenue has a west-east alignment of azimuth 82°, or 262° depending on the direction of view, but the stone rows are not parallel. The northernmost row is aligned 82° 10′, the other 80° 30′; and the avenue terminates at a triangular stone at the eastern end. Lockyer considered that the rising of the Pleiades over the sighting stone would announce the rising of the May Sun – the season of Beltane.[18] One is entitled to ask: why mark the Pleiades when one can simply mark the rising of the Sun itself?

The southern alignment centres upon a small round barrow surrounded by a circle of stones. From the circle, alignments to outlying stones occur at 70° 15′ and 82° 45′. Lockyer believed that these too marked the direction of the rising sun in the May season of Beltane. He therefore derived dates of 1400 BC – 1710 BC for these alignments by astronomical retrocalculation – a remarkable achievement for the early twentieth century. Lockyer thought that the cursus near Stonehenge was similarly aligned towards the rising of the Pleiades.

Perhaps some kind of ritual associated with crop fertility was required at this season. Indeed, the custom of lighting May fires as part of an ancient sacrificial right continued in Scotland and Ireland until as recently as the eighteenth century. The Celtic *Beltane* festival is derived from a Gaelic word implying 'fire' and the name of the Celtic god *Belenos*. The name appears to mean 'bright fire'. A similar festival called *Samhain* marked the waning autumn sun and was held around the first day of November. It persists today in the Halloween festival and the 'Guy Fawkes' bonfires.

Thus, bonfires would be lit on hilltops all over the country on the appropriate night and it would be the responsibility of the local astronomer-priest to determine the correct day of observance. No doubt, local communities evolved slightly different ways to determine the correct day. The heliacal rising of any star can be used to determine the passage of a year and so we might expect to find monuments aligned toward various prominent stars. Perhaps even the correct *time* of day was considered vital. For example the custom on Walpurgis Night, the eve of May Day, was for fires to be lit at the break of day to ward off witches.[19]

Everyone will surely be familiar with the maypole and the May Day fertility rites, in which young girls would dance around the maypole, spiralling in towards it as they interlace their strands around the pole. At this season, it was traditional for young people to collect the spring flowers and to use them to garland the maypole and decorate their homes. The custom of dressing wells with spring flowers continues in Derbyshire to the present day.

We cannot know with certainty what the ancient beliefs were, or how the fertility rites originated; for customs evolve over centuries and have from time to time been suppressed by Christian conversion. The need for such seasonal festivals suggests a concern that the Sun and Moon deities might not always perform as expected. The result for human communities might be death and hardship from crop failure, drought or even floods. The human sacrifice ceremonies of the Maya and Aztecs reveal the extremes to which the human animal will go if we believe that all-powerful deities must be placated.

A Ritual Calendar?

On the Penwith peninsula at the southern tip of Cornwall, lies the Boscawen-un stone circle. Though small by comparison with some of the others, it has an interesting story associated with it. It is mentioned in an old Welsh triad as one of the three principal *Gorsedds* of the island of Britain. A Gorsedd is a committee of Druids occasionally convened to award titles and to plan activities; or at least so its purpose is understood by the modern Welsh bards, who claim to have inherited many genuine ancient traditions. A definition of its ancient purpose might be as a court of justice, or perhaps a ceremonial assembly of local government.

It was Lockyer again who drew our attention to the Gorsedd circle following his address to the Swansea Gorsedd of 1907.[20] It was there that he met with the Reverend John Griffith, a Welsh scholar who had made a study of bardic customs and believed that they preserved ancient Celtic knowledge of astronomy. Influenced by Lockyer's book *Stonehenge*, Griffith had written an article in *Nature*. He drew attention to the similarity in layout of a traditional Gorsedd circle to that of the Boscawen-un and Tregaseal stone circles.[21] When it came his to turn to speak from the central stone, Lockyer delighted the assembled bards by telling them of his recent visit to Boscawen-un and that their Gorsedd was the oldest surviving institution on the planet.

One wonders if Griffith's paper in *Nature* would pass a modern academic referee. He describes the revived Gorsedd circle as a ring of twelve stones placed 30° apart, with a larger central stone. To the east, three outlying stones mark the direction of the equinoxes and the solstices. The stones must be sufficiently spaced for the assembly to stand between them, with the presiding bard at the centre, but the precise number of stones does not appear to have been important. The Boscowen-un circle has a ring of nineteen stones, with additional outliers aligned to the May and November sunrise – critical dates in the agricultural year. Griffith wondered if this detail had perhaps been lost in the transmission of the instructions. He also considered that the 'round table' of Arthurian legend was probably a garbled memory of just such a Gorsedd.

No true Gorsedd has been held in Wales since the English conquest. The original plans for the circle used by the revived modern bards, were among the manuscripts of Edward Williams. He is otherwise known as *Iolo Morgannwg*, author of a collection of ancient Welsh triads and folklore material, known as the *Myvyrian Archaiology* of 1801. He it was, who

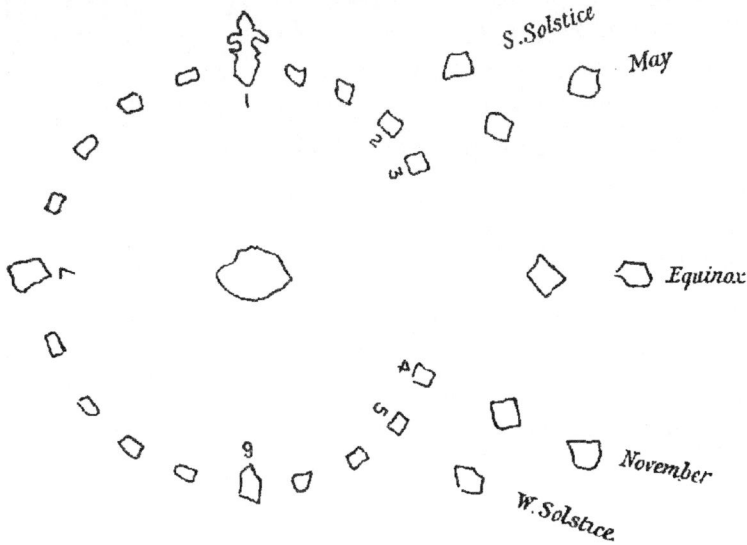

Figure 7.3 The Welsh Gorsedd circle.
(From the Llanover Iolo manuscript, as reproduced in Griffith's article in Nature May 2 1907).

preserved the Third Series of Triads, including those cited in Chapter 1 above. Most scholars think that he invented them, or at best composed them himself from fragments of oral tradition. Yet, as Griffith pointed out, there is no surviving megalithic circle that is *exactly* like the Welsh Gorsedd. It is Iolo's unfortunate reputation that holds us back, but it seems he simply copied and translated some old documents without understanding all their content. Sadly, few of the medieval documents have survived to exonerate him.

Griffith noted that all the plans he had seen contained contradictions between the drawings and the instructions. He therefore obtained a copy of Iolo's original drawing and was pleased to notice that the 'perfect' Gorsedd also showed the additional stones for the May and November alignments, which the modern bards had ignored. The correspondence with the layout of Boscowen-un was striking. Moreover, the alignments were labelled for the solstices, the equinoxes, and also for May and November. If Iolo or some medieval bard had simply modelled the Gorsedd on one of the Cornish circles then they would have had to know that these alignments were real.

As Michell pointed out, this piece of evidence has been staring us all in the face for the best part of a century. Yet, in their zeal to denounce the astronomical properties of the stone circles, the archaeologists failed to take note of the Welsh evidence. In the Gorsedd circle, we have clear evidence for a continuous tradition of ritual and calendrical astronomy dating back to the Neolithic.

The Spiral in the Sky

Near the Cumbrian village of Langwathby lies the circle of Long Meg and her Daughters; one of the oldest and most magnificent of the British stone circles.[22] The Long Meg standing stone is an outlier, positioned some twenty metres south west of the main circle of sixty-five stones, many of which fell over long ago. On the face of Long Meg is a carving of a spiral in approximately three-and-a-half turns, with cup-and-ring marks below.

The spiral form is a common feature of Neolithic art. A very similar spiral with three-and-a-half turns is carved at Knock in south-west Scotland. Another may be found on the Isle of Man, on the so-called Spiral Stone. Within the mound at Knowth, in the Irish Boyne valley there is an exquisitely carved spiral of seven turns, together with another lesser spiral and an enigmatic fan-pattern. Decorative spirals adorn many other stones and artefacts found at Knowth and at the nearby Newgrange mound.

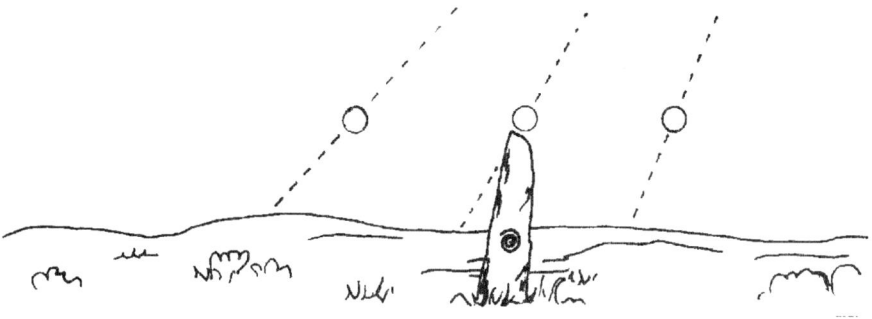

Figure 7.4 Horizon astronomy on a wobbling planet
On a stable world, the sun will rise exactly over the standing stone on the same day each year. However, if the rotation wobbles, then it would oscillate to the north or south of the average position, just as if the observer had travelled north or south to make the same observation.

What could be the significance of this spiral form? Moreover, why does it come to prominence around 3000 BC and during the late Neolithic? Some researchers have viewed the spiral as a symbol of the Earth God; others would see it rather as some kind of solar symbol.[23] Is there any significance to the number of turns? When the spiral motif is used as mere art the number of turns shows no discernible pattern, but when it occurs alone or in a position of prominence, the preferred form is for either three-and-a-half or seven turns. Spirals of similar form dating from this period are also found in other parts of the world.[24]

The astronomical significance of the spiral art at Newgrange and the other Boyne valley tombs was examined by Brennan. He suggested that the spiral form was perhaps a two-dimensional representation of the Sun's motion as it swings from solstice to solstice.[25] Rather, the actual track of the Sun appears to follow a three dimensional helix, which when flattened out *could* be represented by a spiral.

The present author's published hypothesis is that the Neolithic spiral represents a motion of the Earth's axis that is no longer extant. If the crisis that we call the Flood were consequent upon a nutation of the axis and a pole shift, then this wobble would produce abnormal variations of weather and climate. These would demand representation in the calendar and perhaps also in seasonal religious rituals. Does the maypole festival and its spiral dance perhaps celebrate the spiralling of the sky and its uncertain effect upon seasonal fertility?

If this crisis occurred at about 3100 BC then it would also explain why spiral decoration came to prominence during the early third millennium BC. The North Pole would appear to be following an irregular spiral motion about the point that would become the new celestial pole. This pole was itself moving, as discussed in Chapter 4 above, due to the precession of the equinoxes, but this small motion was probably unnoticed by ancient people. A spiralling motion would also betray itself in horizon alignments. The positions of rising and setting would oscillate to the north or south of their average positions for the time of year.

The effect can perhaps be visualised a little better by considering a wobble of the axis, of amplitude say: two degrees. Over the full cycle, the rising of an astronomical body on the horizon would swing a degree further north or south of the average position. An equivalent effect can be reproduced if the observer travelled one-degree north or south to observe the same rising. It would make sense for an ancient astronomer to mark the mid-point by an alignment, as the motion would eventually converge to it. If instead the extremes were marked then they would rapidly become obsolete.

Therefore, it may be that *some* alignments would today be perhaps a degree or so away from our retro-calculated ancient position (which of course assumes no such wobble). One alignment may perhaps preserve an ancient northern extreme, while another gives the southern. This might lead archaeologists to claim that the alignment is poor or non-existent, or even to miss its significance altogether.

The precise period of the nutation is also hard to determine, as will be examined in a later chapter. The period of the modern Chandler wobble is just over fourteen months, or around 432 days. The period of the core-wobble is even less certain, but should be of the same order. A twelve-month period and a fourteen-month period coincide roughly every seven years

$$6 \times 14 \ = \ 7 \times 12 \ = \ 84 \text{ months} \quad = 7 \text{ years}$$
$$6 \times 432 = 2592; \quad 2592 \div 365.25 = 7.096$$

Thus, if the ancient astronomers were aware that the celestial pole was converging on the true pole, then a spiral of seven turns would ideally represent the motion. For horizon astronomers, the period from extreme rising or setting back to the same position would be roughly seven years. From one extreme to the other, or from mid-point back to mid-point, would be three-and-a-half years. A seven-year unit is also evident in many of the ancient Celtic sagas, as if they used it as a convenient long measure of time.[26] The reader may like to consider these coincidences.

The Astronomy of the Druids

There has long been an assumption among historians and archaeologists that the historical Druids mentioned in Roman sources were solely a phenomenon of the late Iron Age and could have no connection with the religion of the pre-Celtic peoples. Equally, popular sentiment would view the stone rings and related monuments as 'Druid circles', with little concern whether the evidence supports this case. So where does the truth lie?

It is true that our best evidence for the activities of the Druids comes principally from late classical writers and associates them with the Celts of Gaul and the British Isles. Caesar tells us that the Druids' system of training was invented in Britain.[27] Of their astronomical knowledge (probably citing Posidonius) he remarks:

> They have much knowledge of the stars and their motion, *of the size of the world and the earth, [author's italics]* of natural philosophy, and the powers and spheres of action of the immortal gods, which they discuss and hand down to their young students.[28]

Celtic knowledge of astronomy is confirmed by the sophistication of the Coligny calendar discussed previously and the inherent conservatism of calendrical science must suggest that such knowledge was very ancient. However, it is not immediately apparent how the stone circles and other alignments could have been used to measure the size of the Earth; and indeed what practical purpose would such knowledge serve? Could the Druids have been engaging in pure science?

Despite the present author's open mind on these subjects, it is perhaps better to regard all ancient astronomical knowledge rather as *precise astrology* with a religious or divinatory motive. As the above quotation itself shows, their true interest lay in "the spheres of action of the immortal gods". Any pure scientific knowledge that they accumulated must be regarded as a by-product of these objectives.

There is no reason, other than a hangover from outmoded diffusionist thinking, that would prevent us regarding ancient Europeans as being at least as capable of divinatory astronomy as were people in India or Babylon. It is almost as if a kind of white racial prejudice were operating in reverse! However, unlike these other societies, the Druids' astronomy was never written down and so no hard records of it have survived. Julius Caesar tells us that the Druids' religion forbade them to commit their teachings to writing, because they did not wish their philosophy to become generally known.[29] In other words, secret knowledge is power, and he who can predict the actions of the gods in the sky might claim to be influencing them.

Although non-classical sources of history are often regarded as somehow less authentic, these too leave a similar impression of Druidic astronomy. A Welsh triad tells us of three great astronomers, who could foretell the omens of the stars until doomsday.[30] Manannan Mac Lir, the Druid after whom the Isle of Man is named, is said to have navigated across the Irish Sea using his knowledge of astronomy.[31] The arcane language that we find in the so-called Prophecies of Merlin also shows us that even abnormal variations of the sea were associated with the malignant power of the stars.

The practical application of astronomy for navigation across the open sea is evident in Plutarch's story of the 'stranger'. In Plutarch's dialogue: *The Face on the Moon* the mysterious stranger claimed to have returned from across the Atlantic Ocean, to visit "the Great Island" – the name that he gave to continental Europe.[32]

Plutarch's dialogue is prime testimony to the high standard of astronomy attained in Celtic Britain. He tells us that at intervals of thirty years when the planet Saturn (the star of Cronus) returned to the constellation of the Bull, the Britons sent out a colonial expedition, consisting of several rowing boats, to an island in the Atlantic Ocean. This island lay in the general direction of the summer sunset (i.e. the north-west) and was the way to a further continent, "the great mainland" beyond. That these voyages took place in the summer months is confirmed by his statement that they would put in at outlying islands, where:

> ...the sun passes out of sight for less than an hour over a period
> of thirty days – and this is night, though it has a darkness that
> is slight and twilight glimmering from the west.[33]

This is a very fair description of the situation in summer just south of the Arctic Circle, in the Shetland Islands, or perhaps the Faeroes. It is self-evident that such voyages would require long-term observation of Saturn and a thirty-year calendar cycle, as previously discussed. The strong evidence for such a calendar cycle, together with the present author's reconstruction of the Coligny calendar, would lend support to Plutarch's statements.

The stranger, according to the dialogue, was one of these colonists, who had served his thirty years and was allowed to return home with the relieving party. He somehow found his way to Carthage, because the god Cronus was so highly honoured among the Carthaginians. There he met with a certain Sextius Sulla, the character who relates the story.

That these ocean voyages were a very ancient custom is testified by the myth that Cronus was believed to be eternally imprisoned upon one of the Atlantic islands, "confined in a deep cave of rock that shines like gold". The expeditions were therefore not some kind of penal colony, but rather a pilgrimage in remembrance of this ancient deity.

It is interesting therefore, that we should have, in the Callanish stone circle on the Isle of Lewis, an example of a Pleiades alignment, together with alignments to the extremes of the summer Sun and Moon. Such ritual alignments are just what would have been required, probably in conjunction with an oral tradition, to preserve the ideal astronomical conditions when the Atlantic voyage should be made. That the colonists also used astronomy to determine the proper time to expect the relieving voyage is again supported by Plutarch's dialogue:

> Here then [*on one of the half-way islands*] the stranger was conveyed, as he said, and while he served the god, became at his leisure acquainted with astronomy, in which he made as much progress as one can by practising geometry...[34]

The arctic summer and winter was a subject of great interest to the classical geographers. Only for the Atlantic coast, did Mediterranean knowledge of the world extend so far north. Another version of this northern geography is offered by Caesar, again probably citing Posidonius:

> ...And it is believed that there are also a number of smaller islands, in which according to some writers there is a month of perpetual darkness at the winter solstice.[35]

This describes conditions north of the Arctic Circle and could only apply to northern Iceland or Norway; but Caesar believed it to describe northern Britain. He verified the shorter nights himself, using a water clock. The horizon-based astronomy of the Britons is further revealed by Geminus, who quotes the lost book of the Greek explorer Pytheas (c.325 BC). Pytheas recorded that the nights were just two hours long at the north of Britain, and remarks, "the barbarians showed us where the sun goes to rest".[36]

That the ultimate source of this knowledge lay with the Druids of Britain is revealed by Pliny in a seemingly innocent geographical error. He evidently understood the spherical nature of the Earth and in discussing the phenomenon of the summer and winter solstices, he remarks on the days of continuous daylight or darkness that must prevail at the poles.[37] However, he comments that continuous days and nights occurred in the island of Thule, which he believed to lie six sailing days north of Britain; and that they also sometimes occurred in the British island of Anglesey.

Thule is another of those ethereal Atlantic islands that we discussed in the first chapter. It seems to have been an amalgam of various northern locations, but it was most likely a report of the Isle of Lewis, to which aspects of Iceland and Norway subsequently became attached.[38] One can understand why reports of Arctic conditions might be associated with these places, but why should Pliny think that they could prevail as far south as Anglesey?

The answer is again supplied by Plutarch's story of the Atlantic voyages. Anglesey was a cult centre of the Druids and a bastion of their resistance to Rome; as is indicated by the importance that the Romans attached to the reduction of the island. The knowledge of the ancient Atlantic voyages must have been among the vast store of orally transmitted wisdom that was obliterated along with the general slaughter of the Anglesey Druids. However, there is every reason to believe that it was never forgotten in Ireland and that Celtic voyages to Iceland continued right up until the age of the Vikings.

Did the Druids remember an ancient time when the Arctic Circle and conditions of polar darkness prevailed as far south as the British Isles? If so then it would support the present author's theory that the Earth has wobbled on its axis in the not-too-remote past. A nutation of more than a few degrees of latitude is surely too great to accept and this could reduce the Arctic Circle no lower than southern Iceland, or perhaps the Shetland Islands. However, we may be sure that even this would produce worrying climatic variations at north British latitudes, severely curtailing subsistence farming and fishing activity.

The recognition that even in ancient times the Britons were a maritime society, who made regular voyages into the Atlantic, supplies the initial stimulus for their precocious interest in astronomy and measurement. A seafaring island society would have to take account of seasonal climate and tidal variations. Herein lies the necessity that is always the mother of invention.

Measuring the Earth

Before Columbus sailed to America, everyone believed that the world was flat – or at least, so we are taught. Even a reading of Pliny's Natural

History, which is but an encyclopaedia of the state of knowledge in the Roman world, will reveal that this was never true.[39] Certainly from the time of Pythagoras onwards, the classical geographers were aware that the known world was a portion of a sphere, and even that the ocean covered the greater part of it.

Why then should we deny that the Druids or the ancient Hyperboreans could also have established this fact of nature? This is a further relic of outmoded diffusionist attitudes and the scholarly prejudice that all wisdom among the northern 'barbarians' must have spread from the civilised Mediterranean.

Pliny cited the variation of the shadow set by a travellers' sundial, in Venice, Rome and Ancona at the same hour of the day, as evidence of the world's curvature. That a similar curvature occurs from east to west was proven by the recorded sightings of eclipses. The solar eclipse of April 30, AD 59 was seen in Italy between 1 PM and 2. PM, but was recorded three hours later in the day by the Roman army in Armenia; and a similar time delay was noted for a lunar eclipse seen by Alexander the Great in 331 BC. Why should we deny that the northern megalith builders were capable of noticing these same phenomena, especially if, as is so often claimed, they were observing and predicting eclipses?

Perhaps the best known deduction of this kind is the attempt by Eratosthenes to calculate the size of the Earth. A Greek astronomer living in Alexandria, he noticed that the shadow cast by a gnomon varied from day to day. At Syene near Aswan, the shadow vanished at noon on the day of the summer solstice and sunlight was observed to penetrate right to the bottom of a well shaft. By measuring the shadow on the same day at Alexandria, he concluded (assuming the rays of the sun to be parallel) that the distance between Syene and Alexandria was one-fiftieth of the Earth's circumference.

Furthermore, Eratosthenes had the distance between Syene and Alexandria paced out and established that it was 5,000 Greek stadia. He therefore concluded that the circumference of the Earth was 250,000 stadia, or about 40,000 kilometres. This is as accurate as could be expected, given his crude estimate of the distance between Syene and Alexandria. However, the logic is flawless.[40]

The size of the Earth could similarly have been deduced by any ancient society capable of measuring the angle of a shadow. We know that the megalith builders of Britain and Gaul were indeed such a society, but we have only Caesar's statement to indicate that the Druids actually did measure the size of the Earth. We cannot know how accurate their estimates were.[41]

Megalithic astronomy was horizon astronomy and the stone circle builders would certainly have known that everywhere the Sun rises due east and sets due west on the days of the equinoxes.[42] On every other day of the year, the rising and setting points vary according to the latitude at which the alignment is constructed. On the days of equinox, the angle of the noon shadow will give the latitude of the site.

Similarly an alignment to the heliacal rising of a star, or to the Pleiades, varies according to latitude and must be established locally. It would not be possible for the Boreads simply to issue a blueprint and for the locals to build their temple to instruction. Astronomical wisdom might be retained within a secret society, but all alignments would have to be surveyed on-site by an astronomer-priest.

This brings us back to the fundamental question of *why* were astronomically aligned monuments built in the years between 3100 BC and 1500 BC; and why did they stop building them? Did something happen around this time that rendered such activity no longer necessary? Certainly, it was not a loss of the basic astronomical knowledge. For, as was discussed above, it survived among the Celtic Druids until first the Romans, and later the Christian monks, pursued them to extinction. A little of this knowledge may even have been retained by the Welsh bards into medieval times, although one doubts whether they understood scarcely a word of it.

Lines in the Landscape

In 1925, Alfred Watkins published his much-cited classic *The Old Straight Track*, which expanded upon his first book *Early British Trackways*. He coined the term *ley line* to describe a straight alignment of ancient sites and settlements over a long distance. He believed that ancient people preferred to travel in straight lines and that to facilitate this, they marked the most direct path between important locations by standing stones, tumuli, or even by planting long-lived trees. Perhaps they even deliberately placed their settlements and henges upon these alignments.

The subject of ley lines and long distance alignments was largely ignored by early twentieth century archaeologists, until it enjoyed a popular revival in the 1960s. The revision of archaeological thinking that commenced in that decade, together with the astro-archaeological work of Thom, Hawkins, and others, revealed that the ancient Britons were perhaps not such barbarians after all. Ley lines came to be associated with UFO's and ancient extraterrestrials – the craze of the day – and with dowsing; with earth-energy grids; and even with lost Atlantis. The archaeologists took a step back!

One aspect of Watkins' original proposals that has stood the test of time is his suggestion that beacon fires were lit on the hilltops as a method of sighting straight leys.[43] Watkins even suggested that the reflection of such a fire in a pond or lake could be employed to ensure absolute straightness. However, even Watkins came to see that straight paths over every hilltop were perhaps, not the most practical routes for people and pack animals laden with goods. So what could be their true purpose?

The existence of beacon hills throughout the British Isles and indeed elsewhere is not in doubt, nor is the fact that many of them have been

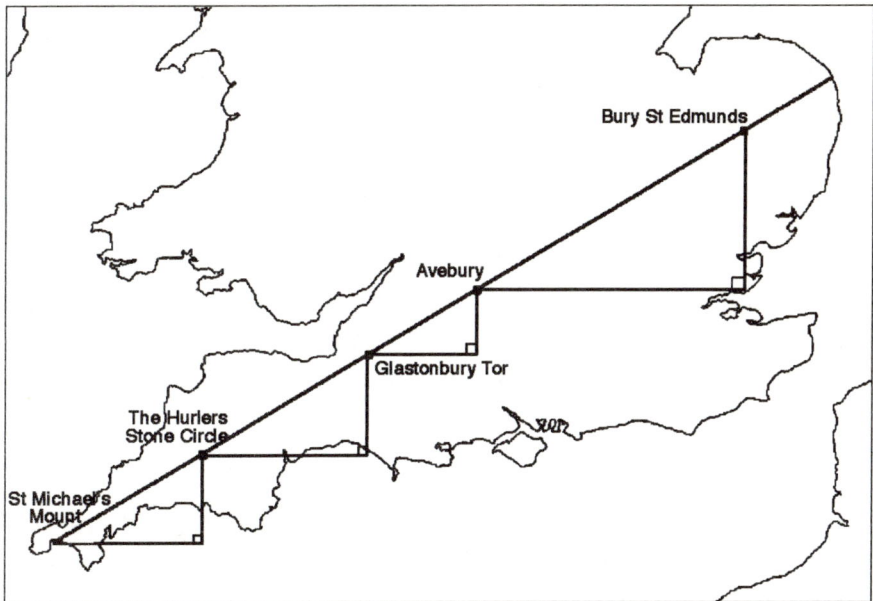

Figure 7.5 The St Michael Ley Line and Longitude
Any series of east-west sighted alignments and triangles could be used in this way to determine longitude for mapping purposes. There is no requirement that the line must be perfectly straight. All that need be measured are the distance between the two corners of each hypotenuse (the line of sight) and its angle with the north-south line. Everything else could then be determined by scale drawing and simple addition.

used since ancient times. Beacon fires were lit as warnings as recently as the Napoleonic wars. By very definition, a beacon fire must be visible from the next beacon and between each beacon hill is a straight line. Whether other sites were then placed upon such 'leys' is a matter beyond our present scope. For those who wish to pursue the subject, there is an abundance of literature.[44]

Although many ley hunters would regard the alignments as dating from as early as the Neolithic, it is usually only possible to claim an alignment if later Bronze Age barrows and Christian churches are also included. It is reasonable to assume that many churches occupy converted pagan sites, as is testified by the presence of ancient stones and long-lived yew trees in so many modern churchyards. Watkins' leys were all straight sighted lines, with each point visible from the next and usually running for no more than ten or twelve miles.

In addition, a number of long distance leys have been from time to time proposed. Perhaps the most significant of these is the Stonehenge ley.[45] This famous line runs nearly north south. Commencing at Frankenbury Camp, it runs north via Clearbury Ring, Salisbury Cathedral, Old Sarum and on to Stonehenge. The line then continues north to Charlton Clumps, a hilltop that is also a modern Ordinance Survey triangulation point. Some would then wish to pick the line up again at Cirencester (although there are no points of significance between) continuing to Nottingham Hill and on to Pershore Abbey.

Critics of such long distance leys will cite the fact that these lines on the map only *appear* straight when a broad pencil is used on a large-scale map. On a closer examination it may be seen that, as in the example above, the best-fit line will run slightly to the east of one site, while passing through the western side of another. This obsession with straightness is shared both by ley enthusiasts and by their detractors alike: *real* ley lines must be straight!

Another example of a long-distance ley line is the so-called St Michael Ley, first proposed by writer John Michell.[46] Several variants of it have subsequently been proposed, but each is as likely or as unlikely as is another. In Michell's analysis, the line links several places associated with dragon legends and dragon-slaying saints. The line commences at St Michael's Mount off Penzance, then inland to the Hurlers stone circle near Liskeard. The next significant points are Burrow Mump and Glastonbury Tor in Somerset, both natural hills with ruined churches on their summits. From there the line proceeds across to Avebury Henge, the church of Ogborne St George and on across country to Bury St Edmunds.

Again, critics would argue that the line is not perfectly straight when subjected to scrutiny.[47] St Michael's Mount and the two hills are natural features and so their alignment can be nothing more than pure chance. As most sites are too far apart to be sighted from the next, there must surely have been other intermediate points. The central portion of the line lies close to the ancient Ridgeway footpath and its continuation in the Icknield Way.

Of course, there are numerous other alignments in the landscape and the over active imagination of a few enthusiasts can weaken an otherwise good case.

The importance of hilltop sites in these alignments brings us back to the custom of the beacon fires. We may reasonably infer that the more recent customs pertaining to the Beltane, Midsummer and Samhain fires are a continuation of ancient practice. We cannot know how or why such customs came about. However, the assembled Druids on the hilltops, as they awaited the light of dawn to set their fires, could not fail to have noticed that fires were lit on the hills to the east of them slightly before dawn. Therefore, they might reasonably conclude that the dawn had already occurred in those places. They might then notice that the fires to the west of them were lit slightly later than their own. Would they not conclude that the Earth's surface was curved from east to west? If one may draw an analogy from evolutionary biology: one might say that the Neolithic Britons were perfectly "pre-adapted" to make such a discovery.

The Druids would already know, from the horizon alignments of their megaliths, that the rising and setting points of the Sun, and the length of the shadow, vary from north to south. Therefore, they were again pre-adapted to make the same discovery that Eratosthenes made. They could similarly have calculated the difference between the angle of the noon shadow cast by a monolith on the day of equinox and thereby derived the angle at the centre of the Earth. They would then have had to measure the north south distance between the two sites just as Eratosthenes did. How would they accomplish this?

The only way to measure such a distance would be to mark out and clear a perfectly straight path between the two points and to then pace it out, or measure it with a rod. This becomes, by default, a 'ley line'. The line does not even need to be exactly north south. If you know the distance between any two points and two angles of a triangle then the length of the other sides can be derived by scale drawing and simple ratio. You don't even need to know the theorem of Pythagoras.

Perhaps, a line once cleared and surveyed in this way might be preserved as a footpath if it were useful to local people. Other lines would disappear as rapidly as they were made. The principle, once established could be used to create a mapping grid, not unlike that used by the modern Ordnance Survey. While this does not prove that ancient Britons covered the landscape with ley lines, it would at least give us a reason why they might want to. If the Druids did indeed measure the size of the Earth, as Caesar informs us that they did, then they must have used a method very similar to that of Eratosthenes.

Eratosthenes' method works perfectly for differences in latitude, because the instant of noon is the same for both ends of the measured line. The ancient Britons may have known that the Earth's surface was curved from east to west, but to establish the precise longitude is a much more difficult task. The Roman geographers were only able to estimate longitude by noting the time of day at which an astronomical event occurs. Lunar eclipses are ideal for this purpose; indeed this was the only way of establishing longitude before the invention of accurate mechanical clocks.

Therefore, how might the Druids have used ley lines to establish their longitude and to measure the dimensions of their island? It is reasonable to assume that they would need their own equivalent of our Greenwich Meridian and that they would have placed this at the centre of the island. Perhaps this was the original purpose of Silbury Hill near Avebury (a mound for which no other purpose has ever been established) to serve as a point of origin for their surveying grid.[48]

If a beacon fire is lit on the east coast at dawn (or better still at moonrise) then it will be seen by the next beacon hill inland. They then light their fire, which is seen by the next beacon, and so on across the country until the message arrives at the toe of Cornwall. If they then measure the time interval until dawn (plus an allowance for the transmission delay) then this gives the time difference between east and west coast. The time interval should be short enough to be measured by a simple water or sand clock.[49] Repeat the procedure several times and take the average. If the circumference of the Earth has been previously calculated then a simple ratio gives the approximate distance between the east and west coasts and a little trigonometry establishes the longitude. None of this science was beyond the capability of ancient people.

Therefore, we have another of those chicken-and-egg riddles. Did the custom of lighting beacon fires originate with a scientific experiment by the Druids to measure the size of the Earth, or did it evolve because they were already lighting fires as part of a religious ritual? Perhaps a little of both is the answer.

Again, we must ask what necessity might have stimulated Late Neolithic people to develop such a precocious interest in astronomy. If the length of day and the tilt of the axis were changed by an extraordinary event at the close of the fourth millennium BC, then our ancestors would have needed to recalibrate their calendar. We have gone some way towards answering the question posed at the beginning of this chapter.

The Druids inherited an antediluvian calendar, which we know as the Calendar of Coligny; and which was calibrated by the number of days in the orbit of Saturn. This ancient knowledge would have told them that Saturn should return to the Pleiades in an expected period of 10,800 days. They could therefore measure that the sky now appeared to be rotating faster by approximately 157½ days in each thirty-year orbit, or about 5¼ days per year.

However, the astronomer-priests would also know that every fifty-six orbits of Saturn, its correspondence with the Moon had *not* changed. This was their absolute time standard: fifty-six average synodic revolutions of Saturn equal 717 lunar months. Therefore, they must have deduced that it was the Earth that revolved and not the sky. They must surely have known that it was the Earth's own revolution that had changed since their lunar calendar was devised.

Something of this arcane knowledge may be preserved for us in that garbled passage in the Prophecies of Merlin and so it may be informative to look again at this:

> The malice of the planet Saturn will pour down like rain, killing mortal men a though with a curved sickle... The Moon's chariot shall run amok in the Zodiac and the Pleiades will burst into tears. None of these will return to the duty expected of it... In the twinkling of an eye the seas shall rise up and the arena of the winds shall be opened once again.

We may recognise in this a fair description of the aftermath of a pole shift; and of a world wobbling on its axis. The ancient Druids must have been terrified that this might all happen again. It is perhaps not surprising therefore, that the classical historian Arrian should record the Celtic boast that they feared no human agency. Their sole concern was that the sky might fall on their heads, or that one-day, the sea might rise-up again to drown them.[50]

References to Chapter Seven

1 Pole Tides are discussed in greater detail in chapter 9 below.

2 See chapter 2 above.

3 Castleden, R, 1987, pp 6-7

4 See, for example, the discussion by Roland-Giot, P, 1994, pp624-6.

5 Burl, A, 1985, p25

6 Castleden, R, 1987, pp 24-27

7 Burl, A., 1993, pp29-30

8 Brothwell, D., and Krzanowski, W., 1974

9 Ray, T.P. 1989, p344

10 Lockyer, J.N. 1909, p 430.

11 Ibid, p 518

12 Ray, T.P. 1989, p 345

13 Nature, 412, 2 December 1999

14 Somerville, B., J. Brit. Astron. Assoc. 23, 83-96 (1912), p 89.

15 Hawkins, G.S., Science, 142, 3654, pp 127-130 (1965).

16 Burl, A., 1993, p 180.

17 If you wonder how important this was then you may like to follow the venerable Bede's correspondence with the Pictish King Nechtan about the proper determination of Easter day. This shows us the opposite problem: that of calculating a lunar festival date under a solar calendar.

18 Lockyer, N, 1909, p 149-155.

19 James, E.O., Seasonal Feasts and Festivals, 1961, pp 312-313

20 This story is related in John Michell's book *A Little History of Astro-Archaeology*. The discussion of the Druids' May-November year may be found in Lockyer (1909) pp 442- 450.

21 Griffith, J. Nature, vol 76, May 1907, pp 9-10

22 Castleden, R. 1987, p 139

23 See North, J., 1996, pp 534-6

24 I have examined the occurrence of the spiral in Egypt and Crete in more detail in Chapter 10 of *The Atlantis Researches*.

25 Brennan, M., The Stars and the Stones, 1983,

26 In Wentz-Evans, *The Faery Faith in Celtic Countries*, 1911, p 49-50 is recorded a delightful folk tale of the 'invisible island', which was said to appear and disappear every seven years between Innismurray and the Irish Mainland. Whoever recorded this piece of Irish blarney could not have known that if the amplitude of the Chandler wobble were to be increased significantly then it would produce just such an effect. Such allegorical evidence strongly suggests a distant recollection of a real event.

27 Caesar, The Gallic War, VI, II, 11

28 Ibid, VI, 14, 6; translation by J.J. Tierney, 1960

29 Ibid, VI, 13, 6

30 The Triad of the Three Happy Astronomers, see Davies, E., 1804, p161

31 In The Yellow Book of Lecain, see the English translation of 1868 by William Skene, pp 77-79.

32 Plutarch, The Face on the Moon, 941-942

33 Ibid, 941; translation by Harold Cherniss and William Helmbold

34 Ibid, 942; translation by Harold Cherniss and William Helmbold

35 Caesar, The Gallic War, V, 12

36 Geminus, quoted on p 132 of Heath, 1932.

37 Pliny, Natural History, II, LXXV, lxxvii

38 See for example the description of Strabo (Geography, 4, 5, 5) based on the lost book of Pytheas, which describes the rugged lifestyle of the inhabitants of Thule.

39 Pliny, Natural History, II, LXVII-LXXIV

40 A similar exercise by Posidonius: Cleomides, De motu circulari, I, 10

41 Authors such as Robin Heath, John Michel and John Neal would like to go further than the present author in their assertion that basic units of measure, principally the English foot, were derived in ancient times from the dimensions of the Earth.

42 In fact, this is something of an approximation. As the Earth's orbit is elliptical, the time from spring to autumn equinox is slightly longer than the second half of the year. For the effect of this on Neolithic alignments, see E.C. Krupp, *In Search of Ancient Astronomies*, 1969, pp 59-60.

43 Watkins, A. 1927, The Ley Hunters Manual, pp 19-20.

44 For the definition of a ley line, see Hodges, M.A. 2001, Society of Ley Hunters Members Questionnaire, Appendix C.

45 See Devereaux, P. and Thompson, I, The Ley Hunter's Companion, for a comprehensive survey of ley lines and associated beliefs.

46 In *The View over Atlantis*, 1969.

47 Williamson. T. & Bellamy, E., 1983, p 149.

48 This suggestion is taken from David Furlong's book *The Keys to the Temple*, p 211 and pp 228-234. He suggests that it was used as a surveying platform for the purpose of marking-out long distance alignments and circles.

49 The time difference between the coast of East Anglia and Penwith is about half an hour, so a signal could probably be sent and returned by bonfire more quickly than the Earth actually rotates. If a signal were sent during the day using polished copper or silver mirrors then it might be sent even quicker. There is no hard evidence for the use of sand or water clocks like the standard Greek *Clepsydra* in Neolithic Europe, but these typically needed to be reset about every six minutes. See Landels, 1978, pp 188-189 and 193-194.

50 Arrian, I, iv. The same story recurs in an ancient Irish saga called *The Cattle Raid of Cooley*.

8

The Neolithic World

The Neolithic world of five thousand years ago was physically very different from the world we know today. The climate of the mid-Holocene, at least at temperate latitudes, was generally warmer; and five thousand years of sea level changes dictate that world shorelines of that time had a somewhat different appearance from the modern map.

Several thousand years earlier still, the world lay in the grip of an ice age that, despite a number of interglacial remissions, had persisted, for more than half a million years. Many now temperate parts of northern Europe and America lay beneath deep ice sheets similar to those that now cover Greenland and Antarctica. Sea levels world-wide were consequently much lower and many of the continental shelves off our modern coasts were exposed as dry land. This at least, is what the specialists tell us. Why are they so certain of this?

Earth's Long-term Rhythms

In 1941, in the sixty-third year of his life, the Yugoslav mathematician Milutin Milankovitch published his theories on solar radiation and applied them to the problem of the ice ages. His work unleashed a furore of debate among geologists but he did not live to see his theories generally accepted.

The Milankovitch theory combined the long-term factors affecting the Earth's orbit as first proposed by Croll; and so allowed geologists to retro-calculate the periods of maximum and minimum solar radiation (and hence temperature) for any given latitude. It predicted the ice age deposits that the geologists should find in the field. In particular, it suggested nine periods of minimum radiation over the past 650,000 years which, if the theory were correct, should correspond to the ages of the major ice sheets. It was the radiocarbon and thorium dating methods that allowed ocean floor deposits to be dated with sufficient accuracy, and so verify the long-term variations that have dominated the Earth's climate.[1]

The longest cycle affecting the world's climate is the orbital eccentricity. Over a cycle of approximately 100,000 years, the perturbations of the planets combine to stretch the Earth's elliptical orbit, from near circular to an eccentricity of about 6%.[2] The perturbation of the orbit therefore causes a real variation in our planet's average distance from the Sun, with corresponding changes in the incident radiation. The present-day orbital eccentricity is about one percent.

The second factor is the 26,000-year precession of the equinoxes discussed previously; this pulls the Earth's oblique axis around in a cone and determines the point in the elliptical orbit at which the equinoxes and solstices occur. Today, the Earth passes the perihelion of its orbit on January 3 and aphelion on July 4. Ten thousand years ago, perihelion coincided with Northern Hemisphere midsummer.

The third long-term factor is a variation of the obliquity due to the pull of Sun, Moon and planets. The tilt of the axis is believed to vary

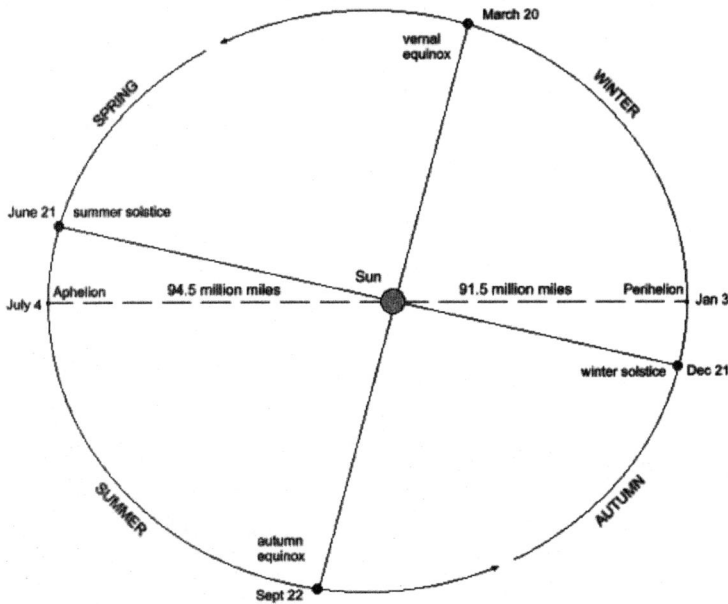

Figure 8.1 Precession and the Earth's orbit.

between 21.8° and 24.4° over a cycle of approximately 41,000 years.[3] Currently the obliquity is 23.44°, with the most recent theoretical maximum occurring about 10,000 years ago.[4] Increased axial tilt should cause colder winters and warmer summers, whereas reduced obliquity would give us a more equable climate.

Milankovitch proposed that the extent of the ice cover was driven by the height of the snow line at each latitude, which was in turn determined by the incident solar radiation consequent upon the three astronomical factors. The combined effect can be plotted as the sum of three sine waves, with the maxima and minima sometimes opposing, sometimes reinforcing each other. The 100,000-year cycle is dominant, followed by the tilt-cycle, with climate changes lagging about eight thousand years behind the astronomical factors that are believed to trigger them.

The definition of 'ice ages' and 'interglacial' periods remains somewhat arbitrary. Geologists define an 'interglacial' as an era when deciduous forests occupied areas of central Europe that had formerly been glaciated. This definition assumes that temperatures elsewhere in the world have followed a similar variation.[5] An interglacial period ends when grassy tundra again replaces the European forests. On this definition, the last interglacial period occurred around 125,000 to 115,000 years ago, and the present interglacial, which we call the *Holocene* or postglacial period, began some 10,000 years ago. All of human civilisation is contained within this geologically short epoch and there can be little doubt that the ice sheets will one-day return. The astronomical theories predict that cooling towards the next major ice age should set in within the next two thousand years.[6]

The ice age theory represents the uniformitarian consensus of modern geology. Catastrophists such as the present author – or at least those who possess a sufficient understanding of the astronomy – must accept the general background of the Milankovitch theory. These long-term periodicities must inexorably continue, regardless of any short-term fluctuations or catastrophes that may intrude.[7] The variations of the Earth's orbit and axis are merely a cause looking for an effect. Once twentieth-century physical science had advanced to the point where glacial deposits could be accurately dated, then the astronomical theory of ice ages was duly proven.

The Climatic Optimum

Few people would be attracted to the painstaking labours of the palynologist. The discipline of wading through peat bogs to extract core samples from the stratified peat is one that most of us will gladly leave to someone else! Yet, climate history is preserved in bog sediments in the form of minute pollen grains. These layers offer evidence of the local vegetation and environment; and the sequence and thickness of each layer tells us about the climate changes that have taken place over the millennia. With the advance of radiocarbon methods, the pollen grains and other vegetable matter can also be directly dated.

The nineteenth century pioneers Blytt and Sernander, working in the peat bogs of Norway and Sweden established a sequence of pollen zones that have remained in parlance among European climatologists for well over a century. Although the Blytt-Sernander classifications have tended towards obsolescence as more work has been done in other parts of the world, they retain their validity for Europe and North America.

A summary of Holocene climate change therefore has to commence with the full ice age (tundra) conditions that prevailed around 18,000 years ago. Climatologists deduce this from the prevalence of the alpine *Dryas* flower, so prominent in the pollen record. Around 12 000 BC warmer conditions allowed trees such as birch and pine to flourish around the shores of the Baltic. However, this seems to have been no more than the beginning of a series of oscillations, for around 10 000 BC the colder conditions returned: this was the so-called *Older Dryas* period.

The true end of the ice age arrived around 9000 BC, with a more general period of warmth called the *Allerod* interstadial, when temperate trees such as oak and elm advanced northwards, but again, this was just a false dawn. A very sudden cold snap called the *Younger Dryas* brought the return of cold conditions to northern Europe for about 500 years. This cold snap was apparently not harsh enough to kill all the established trees and it ended almost as abruptly as it began.

The final onset of warm conditions is dated to 8300–8000 BC with the *Pre-Boreal* zone, which gave way to the *Boreal* period at about 7000 BC. Forests of pine and birch became established far north into Scandinavia and Britain, while a vestigial ice sheet lingered-on in parts of northern Canada. The characteristic of these periods was their relative dryness and a regime of warm summers and cold winters.

As we approach the Neolithic and the time when farming and human civilisations began to expand, we find another transition, this time to warm–wet conditions and mild winters throughout Europe. The climatic optimum began about 6000–5500 BC with the *Atlantic* period. The name derives from the maritime nature of the climate, typical of modern Ireland and Iberia, but which then prevailed throughout northern Europe. Forests of Elm and Lime – a very temperature sensitive species – became prevalent as far north as Scandinavia. In North America, where it is sometimes called the *hypsithermal*, similar warm conditions prevailed and temperate forests became established in formerly glaciated regions.

Despite the fact that this sequence of climate change has been generally accepted for over a century, it is still possible to find archaeological works that make little or no mention of the climatic background in which human activity progressed.

The Neolithic warm period continued up to approximately 3000 BC when another sharp oscillation in the climate brought the onset of warm but dry conditions, similar to those of the Boreal times. The new conditions were therefore defined as the *Sub-Boreal* period. It corresponds to the Late Neolithic period of human activity, when the people of western Europe were building their stone circles and alignments. As we approach more recent times the specialists can also identify shorter oscillations of climate. The short cold snap at the end of the Atlantic period is sometimes termed the *Piora oscillation*, after the Alpine valley where it was first identified. A comparable sharp cooling has also been noted in California at the same era.[8] It was therefore, presumably, a world-wide climate response.

The Sub-Boreal climate of Europe gave way at about 1000–500 BC, without any sharp transition, to modern conditions. This is termed the *Sub-Atlantic* period, because it was essentially wet and influenced by the Atlantic Gulf Stream. No classical source of history extends back beyond the Sub-Atlantic climate transition. Therefore when we look back into legendary and mythical periods, we should appreciate that we see into times when the climate was generally milder than today. However, these variations within Europe should not be over-stressed; the climate of the world may be considered as essentially modern and postglacial since at least 7000 BC (see Appendix C).

As the specialists were able to examine the postglacial climate of many other parts of the world, they quite naturally went out into the field looking for the wet/dry and cool/warm transitions defined by the European pioneers. It is fair to say that they did not find them. However, the end of the ice age and the mid-Holocene climatic optimum certainly did have world-wide counterparts.

When the Deserts Bloomed

In the tropics, the warming at the close of the ice age brought high lake levels in east Africa and Lake Chad, which would indicate a higher rainfall and may point to a strengthening of the tropical monsoon circulation.[9] In the southern hemisphere a similar wet regime is evident in central Australia; and the same transition from a wet climate to a drier one around 2500 BC seems to have brought about the demise of the Indus Valley civilisation.

Increased rainfall may also account for the green Sahara. Before about 3000 BC the present Sahara Desert was a grassy savannah, with scattered lime and eucalyptus trees.[10] Lake Mega-Chad, which throughout most of the ice age had been a vast inland sea comparable with today's Caspian Sea, shrank back to form a lake that was still 40m higher than the present Lake. This may indicate that the tropical rain belt then extended somewhat further north than it does today.[11] Certainly, the rock paintings of the Tassili caves and those along the Red Sea coast indicate the presence of many species, such as hippopotamus, elephant, giraffe and ostrich that are never seen in the art of dynastic Egypt. These animals are now found only in the tropical savannah far to the south.

The River Nile is itself an indicator of mid-Holocene climate. The predynastic cultures of the Nile valley were contemporary with the green Sahara. The Nile delta as we know it today may not even have existed at that time. Back in the ice age, the Nile flowed out into a Mediterranean Sea that stood at a much lower level. The level of the early Holocene Mediterranean stood some 16 m below present sea level and the modern delta has accumulated since the sea level stabilised at its present level, around five thousand years ago. If we may believe classical authors such as Herodotus and Plutarch, the Egyptians remembered the origin of the Nile delta in their mythology.[12] They remembered it not as a gradual rise, but rather as a sudden and catastrophic event associated with the Great Deluge.[13]

Until recently, the only known Neolithic settlement in the Egyptian Delta was that at Merimda on its western edge, dating from 4700 BC.

However, recent drillings have discovered pottery-shards of a similar age some six metres below the water table of the eastern Delta.[14] Many more predynastic settlements may lie so deep beneath the delta sediments that they could never be excavated.

The Dynastic people of Upper Egypt are thought to descend from nomads who sought refuge there after about 5000 BC as the Sahara grew increasingly arid, bringing with them their domesticated sheep and goats. The southward shift of the monsoons and the onset of the desert conditions can be roughly dated by a Neolithic settlement found at Nabta, in the desert west of Lake Nasser. Nearby lie a series of megalithic alignments and a stone circle, small, yet as remarkable as any extant in Western Europe.

The settlements at Nabta were excavated and surveyed in 1997. The remains of cattle and charcoal gave radiocarbon dates around 5,500 years old; indicative that the Nabta basin was home to pastoral nomads during the time of the climatic optimum.[15]

The megalithic structures consist of oval clusters of recumbent slabs. One site gave a date of 4800±80 years BP (2850 BC) and it is perhaps indicative of changing attitudes that the alignments were surveyed for their astronomical significance. Five separate alignments are oriented eastward of the main structure; one at azimuth 90.02° is almost due east, with another at 126°. The three northernmost alignments cluster at 24.3°, 25° and 28°. Northward 1.8° east of north of the main structure lies a monolith.

The nearby stone circle consists of a four-metre circle of slabs, with aligned entrances evidently set out for horizon alignments at 62°; and at 358°, almost north south. The Nabta site lies just south of the Tropic of Cancer and so the sun crosses the zenith on two days around three weeks either side of the summer solstice. The solstice sun around six thousand years ago would have risen at the point on the eastern horizon marked out by one alignment, allowing true north to be sighted along the other.

This parallel development of astronomically aligned monuments in southern Egypt and Atlantic Europe (for no one suggests that the Egyptians needed Druid advisors) is indeed remarkable. Although much smaller, the monument is contemporary with the earliest phases of Avebury and Stonehenge; and the alignments are older than many in southern England. Their purpose was no doubt the same: to mark out the essential days of the calendar for the local community. The question surely is this. What could have driven communities thousands of miles apart to resort to precise horizon astronomy in the years after 3000 BC; and could this reason be the same event that triggered the mid–Holocene climate changes? After 2800 BC the increasingly arid desert conditions rendered the Sahara uninhabitable and, presumably, the pastoral community migrated eastwards to the fertile Nile valley.

Climate Change and the Myth of the Ages

How then, are we to unify the long-term with the short term? If the only factors permitted to drive climate changes are indeed the orbital changes derived by Croll and Milankovitch, then should we not see their influence in the Holocene climate too?

The melting of the northern ice cap is conventionally explained by the warming trend that set in around 18,000 years ago.[16] Based on a combination of the orbital factors it would appear that hotter summers, and hence increased summer melting, was responsible for the disappearance of the ice cover. When summer temperatures peaked around 10,000 years ago, the European and Canadian ice sheets dwindled. However, attempts to explain the Holocene climate transitions solely by the Milankovitch effect have encountered some anomalies.

The decline in summer temperatures that set in since 10,000 years ago should therefore have brought with it a cooling trend. This 'orbital forcing' has been programmed into computerised models, which should be able to predict the Holocene climate in the same way that they account for the long-term ice age variations. However, one recent attempt to model the Boreal climate of 6,000 years ago suggested only a 1.8°C warming at high latitudes due to the orbital factors. The paleoclimatic data suggests that warming was actually closer to 3°C – so factors other than gradual changes to orbit and obliquity seem to have been at work.[17]

One might accept that gradual climate change can be produced by orbital forcing, but it is difficult to see how gradual factors could produce the short-term oscillations at each of the transitions in the Holocene climate. How can gradual changes explain long periods of stability, followed by rapid reconfiguration to another climate mode? However doubtful some of these transitions may be, there appear to have been abrupt shifts, from one climate regime to another, at the end of the Younger Dryas episode around 8000 BC, perhaps also at 5500 BC and again around 3000 BC.

The parallels that may be drawn between this physical evidence of distinct climatic 'episodes' and the various religious and mythological beliefs of a transition from one 'age' to another, are inevitably lost upon most single-subject academics. This is particularly so of those within the scientific disciplines. It is difficult to see how such parallels would ever pass the academic peer-review process to achieve publication in a scientific journal.

The Inconstant Sea

A subject closely related to climate studies is that of Holocene sea level change. This too may seem an arcane matter to most students of religious or historical subjects. However, just as climate change cannot be neglected in the study of human prehistory, neither may we neglect the variations that have occurred around world coastlines.

Again, it is the consequences of the end of the Ice Age that have come to dominate conventional theories of sea level change. Commencing from a position of nineteenth-century common sense, it is apparent that if the Pleistocene ice sheets were as large as geologists believe then vast amounts of water must have been locked up in them. It therefore follows that the sea level world-wide must formerly have stood much lower. A corollary of this is that as the ice-sheets melted (due to the various uniformitarian astronomical factors) then this water was released to flood the low-lying continental margins. Specialists would call this process *glacial eustasy*.[18]

The great rise of world sea level that followed the end of the ice age is known as the Holocene transgression. Its reality cannot be doubted. From a minimum of about 125 metres below modern sea level about 18,000 years ago, the level rose steadily until it stabilised at around 10 metres below present-day sea level around 6,000-7,000 years ago. The remaining rise of the sea has therefore occurred during the period of human civilisation throughout which we have enjoyed apparent stability.

The primary objection to eustatic theories is that sea level change during the Holocene has *not* been uniform, but has occurred rather in a series of episodes. Researchers can find raised beaches and submerged layers of tree-trunks and other land-vegetation in various parts of the world. These indicate that the sea must have stood at a constant level for long periods, before reconfiguring rapidly to a new level. If all sea level changes were gradual then these layers should have been eroded.

This anomaly has been largely overlooked by marine geologists. As each paper cited earlier papers, researchers sought to unify the various episodes as a series of world-wide eustatic variations. Since water may only stand at one level, any beach-feature should have numerous counterparts of the same age elsewhere in the world. Various conflicting models of eustatic control were proposed by oceanographers during the 1960's through to the 1980's but these attempts have now been generally abandoned in favour of a series of regional sea level changes.

Regional variations of sea level imply rather a change in the height and configuration of the continents and ocean basins. World sea level is defined as an irregular oblate spheroid known as the *geoid*. This figure is determined by the gravity at each point on the surface, balanced by the centrifugal effect of the rotation. Local undulations of the solid earth are termed *isostasy*. In other words, the specialists are telling us that there really should be a world-wide eustatic pattern of sea level change – but that this is being masked by isostatic changes at a local level.

The eustatic-isostatic theory remains highly unsatisfactory and no mechanism is proposed to explain these irregular undulations in the shape of the Earth, still less why they should be episodic rather than gradual.

Another cause of sea-level change that has been explored from time to time is that changes to the geoid may be caused by changes in the rotation, or by shifting of the poles. Now it must be stressed that when academic researchers approach this subject, they are not being catastrophists. The changes mooted are always ascribed to *small* and *gradual* processes operating within the interior of the Earth. It should be apparent that even the conventional theories of isostasy must alter the rotational balance and that this would feed back into minor shifts of the poles.

A change in the rate of rotation, as discussed briefly in chapter two, would lead to a migration of the oceans towards the poles (slower rotation or longer day) or towards the equator (faster rotation or shorter day). A geographical shift of the poles would also change the position of the equator. There would always be two neutral points where the new equator crossed the old. The world-wide pattern of sea level change would therefore be one of correspondence in alternate quarter-spheres. In two opposite quarter-spheres the sea-level would fall and in the other two it would rise. The combination of these effects with eustatic-isostatic effects would then explain the apparent randomness of the observed record.

The Swedish geologist Nils-Axel Mörner proposed just such a pattern of 'geoidal-eustasy' for around six thousand years ago, based upon a worldwide survey of evidence.[19] The present author has also proposed such a pattern for five thousand years ago, based upon the apparent withdrawal of the sea in east Asia and South America; and the evidence for Neolithic submergence along the European and Atlantic North American coasts. That these changes may have been catastrophic in nature, rather than gradual and uniformitarian, is suggested by the preservation of myths about the Great Flood, Atlantis, Nu Wa; and the various other non-scientific sources of evidence catalogued in previous chapters.

However, there are good reasons to suggest that this five-thousand-year-old episode was just one of many that the world has experienced. The changes of sea level five thousand years ago are contemporary with the transition from *Atlantic* to *Sub-Boreal* climatic conditions throughout Europe and elsewhere in the Northern Hemisphere. The climate evidence therefore suggests the possibility of earlier catastrophes at 5500 BC and around 8000 BC, at the close of the ice age.

Nothing confirms this better than the recent evidence that has emerged, concerning the rapid changes around the end of the Younger Dryas episode 10,700 years ago. Studies based on Greenland ice cores have revealed a marked change in the ratio of oxygen isotopes *within a period of just twenty years*.[20] Certainly, a rapid retreat of the sea ice at this era seems to have led to a temperature rise of 7°C in southern Greenland, within less than fifty years.

The current explanation for this terminal ice-age event is that it represents the onset of a world-wide circulation of deep ocean water, consequent upon the rapid collapse of the North American ice sheet.[21] An ice-dammed lake at its southern margins in the area of the modern Great Lakes, suddenly collapsed, allowing rapid drainage of the freshwater into the North Atlantic.[22] The rapid rise in temperature then led to further melting of the northern ice cover, with consequent effects upon world sea levels. So here again, we see an apparently catastrophic transition being explained away by a uniformitarian process – the rapid melting of ice.

Perhaps the most interesting piece of evidence here is that the transition from one climate regime to another was completed within a 20-50 year time frame. This also happens to be the decay period of the *Chandler wobble*. If the Earth's rotation were to be disturbed then it must wobble for a period of twenty years before stability could be restored. Even if the sudden collapse of the ice-dam were not directly caused by a pole shift, or the consequent wobble of the axis, then the rapid transfer of such a huge body of water from one place to another must itself excite the Chandler wobble.

Far from proving the uniformitarian case, the Younger Dryas event merely raises a whole new series of doubts.

Volcanoes and Climate

As we have broached the subject of ice cores, we must now pursue this source of evidence in a little more detail.

The ice caps of Greenland and Antarctica were formed as a series of annual layers of compacted snow. In addition to their water content, these layers also trap bubbles of ancient air, pollutants and particles that accumulated on the surface of the snow. The layers therefore lock in a record of ancient volcanic eruptions, which may be examined to give an annual record of the climate.

Volcanoes eject profuse amounts of ash and acid sulphates into the atmosphere, which can circulate all around the world before settling out. The eruptions of Tambora (1815) and Krakatoa (1883) in Indonesia both left their signatures in the ice half a world away at Crête in central Greenland and at Camp Century in north-west Greenland.[23] The annual Ice layers can simply be counted back and wherever similar acid spikes occur, the specialists may infer that a major volcanic event occurred in that year.

The ice-core record identifies a number of these high acidity events, for which the most likely culprit is nearby Mt Hekla on Iceland, as for example, those dated to approximately 1100 BC and 2700 BC. A prominent spike was also recorded for AD 540. Another at 1390±50 BC has been popularly associated with the Aegean explosion of Thera, which may have contributed to the demise of Cretan civilisation; and one around 4400 BC with Mt Mazama, Oregon. However, another high-acidity event dated to 3250±80 BC does not correspond to any known world eruption. A similar high acidity event has also been dated to 5400±120 BC.

Volcanic dust veils are known to have a depressing effect on world climate for several years and this too is evident in the record derived from ancient tree rings. The annual growth rings of trees can be counted back from living trees via a sequence of overlapping samples of ancient wood preserved in bogs. The dendrochronologists have identified low-growth years at dates that correspond very closely to those found in the ice layers. Oak trees from Northern Ireland show that the years with lowest growth correspond to 1153 BC, 3199 BC and 4400 BC, which all lie within the band of uncertainty for the corresponding ice-layer events.

Because of its archaeological significance, particular interest has focused on the identification of the Thera eruption, which tree ring evidence now places at 1628 BC. However, the most interesting feature of this and other low-growth episodes is that the low growth years again congregate into bands of very narrow rings covering a *twenty-year* period. It is difficult to see how a volcanic eruption could affect climate and tree-growth for such a long period.[24] However, due to the statistical uncertainties, it is very difficult to pin-down the precise dates of such events, or to correlate ice-cores and tree-rings at such a distance in prehistory.[25]

Few things crystallise the cautious debate between catastrophist and uniformitarian attitudes in archaeology more succinctly than an exchange that occurred between eminent archaeologists and dendrochronologists in the pages of *Antiquity*. In this case however, the 'catastrophes' under discussion are safely limited to the effects that volcanic eruptions and their corresponding climatic effects have had upon human societies.

Many authors, for example, have sought to link the Thera explosion and its associated tsunami with the demise of Cretan civilisation (which yet others would associate with the Atlantis myth). Some would go further, suggesting that the column of ash from Thera was the 'pillar of fire' described in the Biblical story of the Exodus; and that the Thera tsunami was the inspiration for the Flood of Deucalion. Bronze Age depopulation in Northern Britain has been similarly ascribed to the volcanic activity around 1150 BC.[26] The present author, of course, would venture much further than this cautious standpoint.

Archaeologist Paul Buckland and colleagues would frown upon the interdisciplinary transfer of evidence, stating:

> …there is always the danger that the necessary caution about the ways in which the data are used may be lost. The problem is particularly acute when the events being studied are real, or imagined catastrophes (cf. White & Humphreys 1994).

Catastrophes – be they the destruction of Bronze Age Thera...or the apparent collapse of Middle Bronze Age settlement in upland Britain – are headline news; of such things myths and reputations are born and enter the literature as if proven fact.[27]

The authors were moved to point out that causal links between the archaeological finds and the contemporary climate and volcanic events cannot be proven; and would prefer to see human cultures responding to a series of "socio-economic changes".

The response of dendrochronologist M.G.L. Baillie was swift. He counter-argued that the recent advances in that science have enabled many archaeological events to be dated to a precise year, such that tree-ring dates can almost provide a pseudo-historical chronology for human cultures. His criticism of the archaeologists went as follows:

By challenging...even the existence of the environmental events, they call for preservation of the status quo and seek to shield archaeologists from the unpleasant possibility that there is a catastrophic element in human history. They seem to forget that trees were responding to environmental factors in 1628 BC, 1159 BC and AD 540, not to complex socio-economic models. It is necessary to ascertain the character, the cause and the effects of the environmental factors whether they are induced by volcanoes or cometary dust.[28]

It is a small step from the vulcanism-climate debate to consider the similar effects of a dust veil caused by a comet or asteroid impact, such as that which caused a glow in northern skies after the Tunguska impact of 1908. Would these not also show up in tree-rings and ice-cores? If we find no volcanic eruption corresponding to a low-growth year then we have to consider the alternatives. It may be that an impact event and the subsequent wobble could itself trigger immediate world-wide earthquakes and vulcanism; or there may be a delay of many years in the geological response, such that we may never be able to prove a causal connection.

Again, the concentration of the low growth and frost damage events into twenty-year episodes and the similar short time scales for ice melting at the end of the ice age, raise the possibility of a geophysical cause. The effect of comet and asteroid impacts is not limited to the insertion of dust into the atmosphere. It could also cause a wobble of the rotation, or even a permanent pole shift. These too, would have an effect on world climate and sea levels. We may dispute the amplitude and extent of such events, but modern geophysicists certainly understand how the Earth would behave in such circumstances. It is a great pity that so few of them express their understanding in simple terms; and that so much of the subject remains shrouded in dense mathematics.

The Black Sea Flood

In 1998, the American geologists William Ryan and Walter Pitman caused something of a stir with their book *Noah's Flood*. The gist of their theory is that the Sumerian Flood story in the Gilgamesh Epic recalls a catastrophic flooding of the basin of the Black Sea. This supposedly occurred around 7500 years ago, when the waters of the Mediterranean Sea first broke through the Bosporus Strait.

Ryan was co-chief scientist of the Deep Sea Drilling Project aboard the vessel *Glomar Challenger*, which investigated the sedimentary layers of the world's oceans during the 1970s. It was this expedition, which first revealed that the bed of the Mediterranean Sea had once been a dry desert. What is now the valley of the River Nile had then become a deep gorge, when the pre-Nile cut down into the sediments as the Mediterranean basin gradually dried up. However, these events occurred around *five million* years ago and were followed by a dramatic flooding of this basin when the waters of the Atlantic burst through the opening at the Straits of Gibraltar as a gigantic waterfall. The evidence of tiny deep-sea crustaceans, found in a thin layer at the transition, reveal that the entire basin of the Mediterranean must have refilled within a period shorter than a human lifetime.

When Ryan and his long-time colleague Walter Pitman came to ponder the problem of Noah's flood, they sensibly concluded that the Mediterranean catastrophe was far too ancient to be the source of any human myth. Their attention therefore focused upon the Black Sea, which their contacts with Bulgarian and Russian researchers had revealed to be a much more recent candidate.[29] Research carried out by the Bulgarian academy of Sciences had shown that at the close of the Ice Age, the Black Sea stood some one hundred metres below its present-day level. Moreover, at that time, it had been, not an arm of the sea, but a trapped freshwater lake filled by melt-water from the shrinking northern ice sheets.

Again, it was dead sea-creatures that gave away the date when the Euxine Lake became the Black Sea. Core samples showed an abrupt transition from layers containing freshwater molluscs to those with marine species. Radiocarbon dates proved, to within a margin of error of about fifty years, that this transition occurred around 5500 BC.[30]

Now it will be apparent that this date corresponds closely to the abrupt climate change that European climatologists have long referred to as the *Boreal-Atlantic transition*. Around this time, the climate throughout Europe and elsewhere in the Northern Hemisphere changed from warm-dry conditions to a warm-wet regime. One is entitled to ask if there is a connection between these events; and what might this connection be?

Ryan & Pitman suggest that the submerged shores of the ancient Euxine Lake were formerly home to many races, who migrated westward into Europe and southwards into Mesopotamia, taking the memory of this flood with them. This strikes a chord with the stories in the Welsh Triads, which claim that the ancient Cymry migrated to the British Isles after the Flood, from the lands around the Bosphorus.[31] These triads have long been dismissed as nineteenth century inventions, but this new research surely confers upon them a certain degree of authenticity.

Ryan and Pitman's theory was soon seized upon by the popular press as proof of the Biblical Flood, but did they really say anything new? The changing levels of the Black Sea have certainly been known to climatologists since at least the 1960s, being discussed in various papers.[32] The difference seems to be that their theory represents a uniquely *uniformitarian* explanation for the Great Flood, proposed by two most respectable and eminent American academics. To accept their mechanism one need know nothing of the complexities of astronomy, nor even that the Earth rotates on its axis. Melting ice and Anatolian earthquakes will suffice to explain the Flood.

Yet, many anomalies remain that cannot be explained by the Black Sea Flood theory alone. The Biblical Flood is a deluge from above and Ryan and Pitman's theory does not explain the torrents of rain. Although the theory plausibly explains how a flood story could radiate to Europe and Mesopotamia, it cannot explain how such stories came to spread as far afield as China, South America and Australia (though apparently not to Africa). It cannot explain the apparent memory of changes in the heavens that we find in the Chinese myths, nor the 'giants' and 'dragons' in the sky that we find associated with so many other world flood myths.

Another anomaly comes with the supposed dating of the Mesopotamian Flood to an event as early as 5500 BC. All the other evidence that has been surveyed in this book, be it conventional or unique to the present author, has converged upon a date for the Flood around 3100 BC. Again, the astute reader will have noticed that this later date agrees with another climate transition – the so-called *Atlantic/Sub-Boreal transition*. This would concur with the present author's theory that *all* flood myths are composite myths, merging aspects of both earlier and later catastrophes. It may be that the stories of Noah's Ark, Deucalion and the Welsh Nvydd do indeed recall the Black Sea Flood, but other astronomical and weather phenomena perhaps rightly correspond to an earlier or later event.

Further questions can be posed, which are perhaps rather more scientific. The era of the Euxine Lake corresponds to the period when the North Sea was dry and Britain was linked to continental Europe. Ireland and Britain may also have been linked by a land-bridge at that same era. The submergence of the North Sea land barrier around 5500 BC may be sufficient to explain the transition from the warm-dry climate that had prevailed in Northern Europe, to the warm-maritime conditions of the Atlantic Period. Moreover, is there a connection with the acid-spike found in the Greenland ice at around the same era? It is not explained how the collapse of the Bosporus Strait alone could have triggered other sea level and climate changes in the wider world.

As with so many other academic hypotheses, the Black Sea flood theory is found wanting when its arguments are extended beyond the chosen subject area, or into fields of expertise beyond the excellence of the specialists who have proposed it.

Circular Arguments

It may be seen that the very definition of an ice age is one of circularity. The advances and retreats of tundra within Europe defined the ice age time frame; and as these authorities were cited in later works, the rest of the world was forced into the mould.

Moreover, assumptions about the former Scandinavian ice cap and its melting imply similar ice cover over other parts of the Northern hemisphere, notably Canada and Alaska. The complete absence of ice cover in Siberia – today the coldest part of the world – throughout the entire Pleistocene epoch, must remain an enigma. If the summer sun rises to the same height in the sky in both Siberia and Canada then why

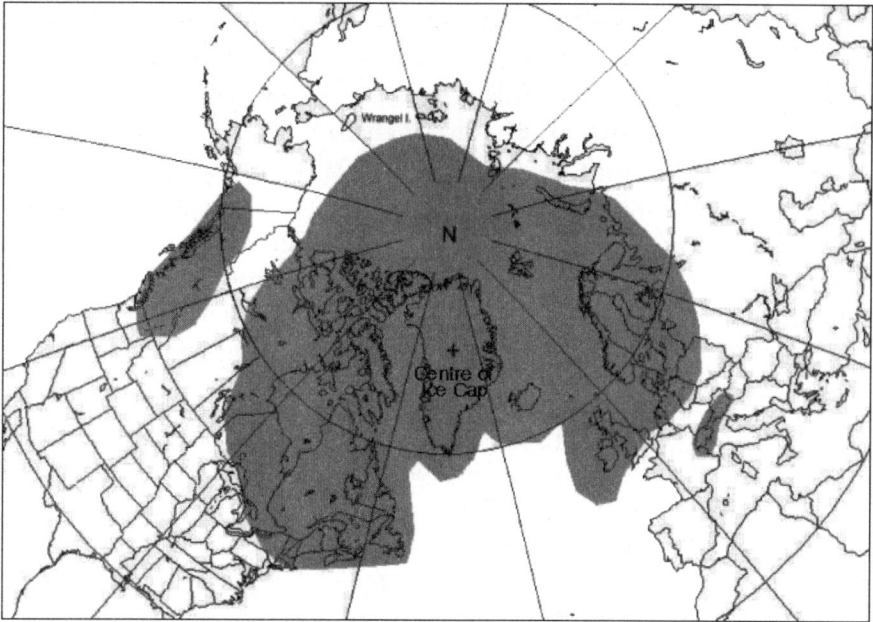

Figure 8.2 The lop-sided northern Ice Cap.
The northern-hemisphere ice cap was apparently centred upon Greenland, while in Siberia,
mammoths munched on daisies and survived on Wrangel Island until as recently as 2500 BC.

was it apparently able to melt the winter snow in Siberia, such that an ice sheet was never able to form? Yet, in Greenland and Canada the snow lingered year after year. The lop-sided northern ice cap is another instance in which common sense has been suspended by uniformitarian theorists.

There is no way to measure directly the thickness of ice sheets that have melted away. Their presence has to be inferred from the rebound of the land in areas such as Scandinavia and Hudson Bay, where the burden of ice has been removed. This apparent rebound is itself estimated from the height and age of the raised beaches in these areas, relative to the assumed world sea level at that era. Yet, these sea levels are themselves derived from eustatic assumptions about how much ice has melted. Like the World Serpent of mythology, the uniformitarian hypothesis endlessly chases its own tail.

The growth and retreat of the ice cover are supposedly driven by climate changes, yet climate and weather are themselves a function of sea level and ocean currents. Since the sea level stabilised close to its present level some six thousand years ago, the whole of human civilisation has arisen. A continuation of the melting during Neolithic times might explain the submergence of the North Sea plain and submerged peat and forest deposits around Atlantic coasts; but how can it explain the *emergence* of the five-thousand-year-old raised beaches of South America, or the similar retreats of the sea from East Asian coasts?

There is nothing in the pattern of evidence that would prohibit a pole shift, or a series of them consequent upon an astronomical cause, so long as one does not insist that they are the *only* process at work. This is where previous catastrophist theories have fallen down. It may be that the quarter-sphere pattern of earlier pole shifts has been submerged by the Holocene transgression such that we may now find evidence of only the most recent episode that occurred around 3000 BC. The arguments against this possibility come principally from physics and astronomy, not from geology or climate evidence.

We may also see evidence for rapid changes that bear the signature of the twenty-year Chandler wobble. Did rapid melting of ice cause a wobble of the axis of rotation? Alternatively, was melting triggered by such a nutation? In addition to bringing unstable weather and climate to the polar latitudes, the sea-ice would be invaded by pole tides; and earthquakes would trigger underwater mudslides and tsunamis. These processes can explain the episodic nature of climate and sea level variation that conventional theories conveniently ignore.

Radiocarbon and 'Wiggle Matching'

There is yet another source of evidence for episodes of change during human prehistory and it comes from the very method that archaeologists use to date the past.

The radiocarbon dating method emerged after the Second World War as a by-product of the Willard F. Libby's nuclear research. Although initially the physicists estimated a slightly incorrect half-life for carbon-14, the amount of radiocarbon remaining in an ancient piece of wood or bone provided archaeologists with the first means to independently date an ancient site.[33]

The method relies on the fact that all living things absorb radioactive carbon-14 in the same proportion that it is present in the atmosphere. When a creature dies, the proportion diminishes exponentially as the carbon-14 decays away. Given the half-life of carbon-14, the age of the sample can therefore be readily estimated. However, as radioactive decay is truly random, there always remains an element of statistical inaccuracy built into a radiocarbon date, which increases with its age. A two-thousand–year–old radiocarbon date should lie within about 75 years of the true calendar date, whereas at fifty thousand years – at the limit of its range – the band of uncertainty widens to around 1000 years.

The introduction of radiocarbon dates during the decade of the 1950's caused a furore in archaeological circles, particularly among Egyptologists. Hitherto, they had used cross-dating methods to establish their chronology, which relied upon the Egyptian and Mesopotamian king lists. They began to find that dates for Middle and Old Kingdom Egypt were coming out consistently younger than expected and consequently, a further twenty years would elapse before Egyptologists fully accepted the radiocarbon method of dating.

Egyptologists were fully justified in rejecting the dates supplied by radiocarbon. For example, the dates for New Kingdom artefacts from Amarna and Tutankhamun's tomb gave dates around 1050 BC, whereas the accepted chronology derived from the king lists placed them 200 years earlier. Moreover, dates for the Old Kingdom pyramid age were seen to be as much as 500 years too young. For the First Dynasty the error widened to 800 years. To accept these dates would require hundreds of years of Egyptian history to be simply written-off.

The dispute was resolved in 1969 when the Nobel Symposium in Uppsala, Sweden formally accepted that radiocarbon dates could be calibrated by dates taken from the growth rings of long-lived Californian Bristle-cone Pine tree. These proved that before about 1000 BC, the ratio of carbon–14 in the atmosphere had indeed been higher and that this excess had caused the uncalibrated dates to appear too young. The outcome was a complete revision of archaeological theories about Neolithic Europe and the Mediterranean; and cultures in the rest of the world could finally be compared with their contemporaries in Europe and the Middle East.

Unfortunately, this was far from being the end of the debate. Although calibration brought radiocarbon and historical chronologies into closer alignment, it still could not provide precise calendar dates. Tree-ring calibration showed the radiocarbon 'curve' to be not a smooth exponential decay, but a sinuous curve with numerous 'wiggles' and undulations. In some cases, this means that a calibrated result could fall at two or three places on the curve – a process known as 'wiggle-matching'. Archaeologists now prefer to take the average of several dated samples.

Moreover, some Egyptologists continued to dispute that a Californian series of tree rings could apply for Egypt. In 1976, the Egyptologist Richard Long showed that Egyptian calibrated dates were now consistently older than the historical chronology, requiring him to 'stretch' Egyptian history. Since then, further work has been done on bog-oaks from Ireland and Germany; and it seems that new research on trees from Turkey may finally resolve this dispute.[34]

However, the discrepancy in radiocarbon dates between 1000 and about 3000 BC is undoubtedly real. Between approximately 1000 BC and the present, the radiocarbon clock seems to confirm the ratio of carbon-12 to carbon-14 in the modern atmosphere.[35] Since the laws of biology and radioactive decay cannot have changed the only conclusion must be that before 1000 BC there was *more* carbon-14 in the atmosphere. Moreover, this proportion must have steadily *decreased* between about 3000 BC and 1000 BC.

Carbon-14 production in the atmosphere is a consequence of interaction between cosmic ray particles (neutrons) and the nitrogen atoms in the upper atmosphere. These are transmuted into carbon-14 atoms, which are almost immediately bound up in the form of carbon dioxide. This is taken-up by plants as they grow. The interaction is therefore also a function of the Earth's magnetic field strength. A stronger field means a lower rate of carbon-14 production and vice-versa.[36] It also depends on latitude. Cosmic ray intensity is strongest at the poles and weakest at the Equator; and the mixing is dependent upon the weather and atmospheric circulation.

While the short-term 'wiggles' in the radiocarbon curve might be explained by irregular changes in the Sun's magnetism, the cause of the long-term variation, before 1000 BC, remains an enigma. Was the magnetic field weaker? Or, was the cosmic ray flux stronger?

The source of the terrestrial magnetic field lies in the rotation of the fluid nickel-iron core. It should be apparent that any pole-shift or change to the length of day, of the kind suggested here and by other authors, would also affect the intensity of the magnetic field. However, the invariability of the Earth's rotation, at least to any significant extent, is taken as a given constant in most academic studies of paleomagnetism.

The present author's theory is that the cause of the discrepancy in uncalibrated radiocarbon dates lies in a pole shift and change to the length of the day that occurred around 3100 BC.[37] If the length of day was formerly only 360 days per year then this slower rotation would have generated a weaker magnetic field and hence more carbon-14 in the atmosphere.

Furthermore, if the dynamics of an impact on the surface caused the outer shell to rotate faster then this change could not be immediately communicated through the fluid outer-core. The inner core would continue to rotate at the slower rate; with the two rates gradually equalising as the Earth wobbled on its axis. The ratio of carbon-14 in the atmosphere would then steadily diminish as the excess was taken up by the biosphere. The period of change between c3000 BC and c1000 BC corresponds closely to the Sub-Boreal period of climate change; and with the period when Neolithic people were building their astronomically aligned monuments.

A Cure for Insomnia?

Have you ever felt that sometimes there are just not enough hours in a day? We have surely all had this feeling at some time, when tiredness prevents us from completing our essential daily tasks. Many of us also have difficulty sleeping at night – especially those who are wont to ponder upon the nature of cosmic catastrophes!

The human brain, and indeed that of other living creatures, has an internal body clock known as a circadian rhythm. The body clock uses alternating periods of light and dark to reset itself so that we all know when we should be awake and when to sleep. However if these stimuli are withdrawn, for example when people are trapped underground for long periods, then the body clock may begin to drift out of synchronisation. Disorders of the body clock may also cause insomnia, as can some habits of modern living, such as so-called jet lag, space travel, or even irregular shift-work patterns. Medical researchers have discovered that just 30 minutes of immersion in light is sufficient to reset the body clock.

However, when left to run freely the circadian rhythm is not, as we might expect, one of 24-hours. The rhythm in the brain is actually a period of 24.3 hours.

We may well ask *why* the human body has evolved such a clock and what is the *advantage* of a 24.3-hour rhythm over one of 24 hours; and why has the brain evolved a reset mechanism? Evolution is a process that operates over millions of years of gradual adaptation and nothing in nature exists without a reason. The existence of such a circadian rhythm may be telling us two things. Firstly, the evolution of a reset mechanism tells us that our ancestors had to be able to survive through periods when the length of day changed. Secondly, it suggests that: until quite recently in evolutionary terms, the length of the day must have been 24.3 hours.

Do a simple calculation:

$$365.25 \times 24 / 24.3 = 360.74$$

Does this biological imperative perhaps tell us that throughout most of recent evolution the human body adapted to a world where there were 360¾ days in a year? Could it be that the change to a year with 365¼ days has happened so recently that life on Earth has not yet had time to adapt? Could this change have occurred at the end of the ice age, or even as recently as 3100 BC?

With Eyes Closed

Prejudice alone prevents specialists from seeing that rapid geological changes cannot be explained by gradual causes alone. How rapid does a change have to be before it becomes a catastrophe? If, at regular intervals of a few thousand years, the Earth is struck by a comet, asteroid, or some other unknown force of cosmic origin, then is this any longer to be defined as catastrophism? Does it not simply become a part of the uniform record of geology? Geologists since Lyell have divided the past into a series of epochs each lasting *millions* of years. Did the Earth perhaps commence a new geological epoch just a few *thousand* years ago?

For too long, scientists have treated our world as if it were somehow isolated and protected from the awesome processes that shape the rest of the universe. Geologists are only just beginning to realise that impact hazards are now a part of their subject, and not merely the province of astronomers.[38] It took a comet collision with Jupiter, live on our television screens, to awaken many to the fact that impact craters don't just occur on the Moon!

However, the suggestion that these regular catastrophic interventions from space might also affect the Earth's rotation and reconfigure world climate and sea levels at regular intervals, remains a step too far. Surely, the argument goes, there would be some hard evidence. Not so long ago, scientific opinion was equally dismissive towards the giant impact that closed the age of the dinosaurs. Yet, once the crater in Yucatan was identified it all seemed so obvious. So where is 'the Neolithic crater'?

Therefore, when we see a consistent pattern in recent prehistory that *suggests* a change in the length of the day, a pole shift and a change in the obliquity, why does this possibility continue to be ignored? The answer is in part historical. It follows from the way that early religious catastrophism was defeated by nineteenth century science and the way that scientific enquiry has become ever more specialised and partitioned.

So let us turn the question around! Prove that the length of the day has *not* been changed in recent millennia. Prove that the episodes of Holocene climate and sea-level change were *not* caused by pole shifts and changes of axial tilt. It is impossible to prove such a negative.

A critic might naively submit that if a comet impact had indeed "knocked the Earth off its axis" then the rotation would wobble, in one mode for about 20 years, in another for about 2,500 years before it regained balance. This must cause pole tides, river floods and lake-bursts. If the poles had truly shifted then we should find a pattern of recent raised beaches and submerged forests all around world coastlines; and at the same era the climate at every latitude would reconfigure to a new mode. The intensity of the Earth's magnetic field would change. If the number of days in the year had indeed changed then surely we would find evidence of it in the construction of the world's earliest calendars? Surely, such an event would be remembered world-wide in our oldest myths and legends? Are these not the very matters we have discussed in the preceding chapters?

Yet there remains a further strong objection to the theory – and it is a very big one indeed!

References to Chapter Eight

1   Hayes, Imbrie and Shackleton, 1976

2   More precisely, an irregular period that averages 96,600 years. A team of French scientists have estimated that without the stabilising effect of the Moon's tidal pull on the Earth, the obliquity would have varied widely throughout geological time, by as much as 85 degrees of latitude. This would prohibit most of the more bizarre theories of recent lunar capture as the higher life forms probably could not have evolved under such an extreme climatic regime. On this, see Laskar, J. et al, 1993.

3   This secular variation assumes, of course, that no factors other than planetary perturbation have affected the obliquity.

4   Lamb, HH, Volume 1 of Climate, Past, Present & Future, 1972, p 31

5   It should be apparent that this would *not* be the case if the ice ages were influenced by a pole shift, as the variations would then correspond only in alternate quarter spheres.

6   Imbre and Imbre, 1979, p178.

7   See the analysis by Paillard, 1998, p378-81, "an especially warm interglacial 400kyr ago, occurred at a time when insolation variation was minimal".

8   Lamb, H.H. 1995, p 142.

9   Roberts, N., 1989, p89

10 I note that Hubert Lamb, in the second edition of his work *Climate History and the Modern World* (1995) discusses some of the same effects of the climatic transitions upon ancient civilisations, that I noted in *The Atlantis Researches* (also published in 1995). Lamb's extensive earlier publications were very influential to the present author's conclusions, although he of course approaches the problem from a conventional rather than a catastrophist outlook.

11 Ritchie, J.C. & Haynes. C.V. Nature, 330, 1987, pp645-647.

12 Herodotus II,13-17.

13 Plutarch, Isis and Osiris, 367.

14 Holmes, D.L., Nature, 363, pp 402-3 (1993)

15 Malville, McKim et al, Nature, 393, pp 488-491 (1998)

16 Kutzbach and Street-Perrot, 1985. (see p 91 of Roberts, N, 1989).

17 Foley, et al, Nature, 371, 1994, p 53.

18 I have discussed this problem in greater detail in Chapter Six of The Atlantis Researches.

19 Mörner, N.-A (1980)

20 Dansgaard, et al, Nature (1980), 339, pp532-3.

21 See, for example, the discussion in Manabe and Stouffer, 1995.

22 Barber et al, 1999, pp 244-8.

23 Hammer, et al, Nature (1980), 288, pp230-235.

24 For example, around 1150 BC, low growth commenced at 1159 BC and continued until 1140 BC. See Baillie, M.G.L. & Munro, M.A.R, Nature 332, pp344-346.

25 To quote the foremost authority: "Even when great quantities of 14C dates are available, the statistical scatter and laboratory bias, coupled with calibration problems, makes it virtually impossible to closely estimate true age to better than a century". See Ballie, M.G.L. (1994), p704.

26 Burgess, C. 1989, Current Archaeology 117, vol X. No.10. 325-9.

27 Buckland, P.C., Dugmore, A.J. & Edwards, K.J., Antiquity 71 (1997) 581-93.

28 Baillie, M.G.L., Antiquity 72 (1998) 425-7.

29 Noah's Flood, p118.

30 Noah's Flood, p149.

31 See chapter 1 above.

32 See, for example, the summary in Lamb (1977), vol 2, pp368-9

33 Libby originally used a half-life of 5,568 years, but later work showed 5,730 years to be closer.

34 See, for example, the discussion in Renfrew. C. 1996.

35 There is also a small inaccuracy caused by the presence of the carbon-13 isotope.

36 Aitken, M.J. *Physics and Archaeology*, 1974, pp 31-32; a variation of 1% in carbon-14 concentration causes an age-error of 83 years, ibid. p 68.

37 See chapter 9 of *The Atlantis Researches*, p 124.

38 Jewitt, D, 2000 pp145-8.

9

Squaring the Circle

We have now established a pattern of evidence drawn from ancient sources, mythology and physical evidence, which suggests that the Earth's axis was disturbed during Neolithic times, around 3100 BC. The world-wide occurrence of Flood myths may be a distant memory of a Neolithic event, or more likely, a composite picture of this and even earlier episodes. The poles shifted geographically and the tilt of the axis in space was also affected; and calendrical evidence suggests a change in the length of the day from 360 to 365¼ days per year.

Despite the consistent pattern of evidence that such a change occurred five thousand years ago, there remains a very substantial objection to the present author's theory that up to this point has been sidestepped. It is considered physically impossible!

Most geophysicists will concur that the rotational poles have remained sensibly coincident with their present positions throughout the vastness of geological time. Moreover, while some very small movement of the poles on the Earth's surface may be permissible, a sudden change in the obliquity would require a quite vast external force to be exerted. Such a force would require a geologically recent collision with a huge comet or asteroid, for which there is simply no hard evidence. The strongest argument against it is that such a massive impact must surely destroy all life on the planet – and yet demonstrably we are still here! How is this circle to be squared?

Some Nineteenth Century Theories

In the mid-nineteenth century, geologists were just becoming accustomed to the reality of past ice ages. For many geologists of the time, there remained a need to find a scientific basis for the Biblical Flood and to explain the ice age deposits of the Drift. One popular idea was that the huge polar ice caps that had supposedly extended down to the Alps had never really existed at all. Rather, the poles had simply wandered around on the Earth's surface, carrying the ice caps with them. This theory had the advantage that it might also explain many other anomalous geological phenomena, such as why fossils of tropical species can today be found even in the polar and temperate regions.

One such theorist was Sir Henry James who, in the British journal *The Athenaeum* of 1860, proposed that recent climate changes had been brought about by a migration of the axis of rotation. This movement, he speculated, had been brought about by the rapid elevation of the world's mountain chains, which had disturbed the rotational balance.

In a reply in the same journal, the Astronomer Royal, Sir George Airey, considered the matter with a greater understanding of the physics.[1] He concurred that this mechanism was indeed feasible, but that the forces required to bring it about were simply too great.

Airey drew upon the work of the mathematician Euler, who had described the wandering of the axis of rotation within a rigid body of the Earth's mass. Even assuming such ideal conditions, Airey concluded that the elevation of a mountain chain, by some unknown explosive process, would cause the Earth's principal axis to migrate. After a sensible interval of wobbling as described by Euler (the true Chandler period being then unknown) the rotation would settle to the new principal axis. He initially considered the sudden elevation of a mountain-mass equivalent to the Earth's equatorial bulge: some 25,000 miles long, 6,000 miles wide and 13 miles high, saying:

> Even if a mountain mass contained 1/1000 part of this matter (which, I apprehend is very far above the fact), the shift of the earth's pole would be only two or three miles; and this, though it would greatly surprise astronomers, and might sensibly affect the depth of waters in harbours, would produce no such changes of climate as it is desired to explain.

Furthermore, Airey went on to argue that: as the Earth is pliable, the first day's rotation would cause the entire planet to deform such that the axis immediately realigns about the rotation axis. He concluded: "This, as I conceive puts an end to all supposition of change of axis". He also ruled out any possibility of a change in the obliquity of the ecliptic:

When the position of the axis of rotation within the Earth is sensibly permanent, its position (excluding the effects of external actions) is also sensibly permanent in the heavens.

Another response, immediately following Airey's contribution in the same journal, pointed out that if the Earth were completely fluid, it would rotate stably about its shorter axis forever and must have done so since its formation.

A few years later, Sir John Evans, president of the Geological Society, in an address to the society, invited mathematicians to prove that geological upheaval and subsidence had brought about a migration of the axis; such that during the ice-ages the North Pole had lain some 15-20° further south, between Greenland and Spitzbergen.[2]

Evans' musings triggered a number of responses. One contribution by the eminent Professor Twisden concluded that the proposed elevation of land could produce a polar wander of no more than 10' of arc, certainly not the 15-20° desired by the geologists.[3] He further alluded to the Euler-wobble: pointing out that the separation of the rotation axis from the axis of figure would produce two vast waves that would sweep round the world every 150 days and submerge the equator to a depth of six miles or more.[4]

Another respondent, the Reverend Hill, tutor of St John's College, Cambridge, pointed out that geologists did not seem to comprehend the vast forces that would be needed; and that a body of the size of the Earth could not be set to rotate about another axis except by a quite substantial external force.

> Internal changes cannot alter the axis, only the distribution of the matter and motion about it. If the mass began to revolve about a new axis, every particle would begin to move in a new direction. What is there to cause this? When a cannon ball strikes an iron plate obliquely, the shock may deflect it into a new direction. The Earth's equator is moving faster than a cannon ball. Where is the force that could deflect every portion of it and every portion of the earth into new directions of motion?

Not only did he rule out polar wander due to internal forces, neither could he conceive of any external force upon the Earth that could alter the obliquity.

Thus, we may see that very little in the present author's arguments is new. It was all considered by nineteenth century science – and completely dismissed!

Sir George Darwin

Enter Sir George Darwin, son of Charles Darwin, the pioneer of evolution. Unlike his rather better known father, he made his name in the field of physical geography. His response to Evans' challenge applied a set of equations worked out by the French mathematician Liouville, which more accurately modelled the rotating Earth as a fluid rather than as a rigid body.[5]

Again, he concluded that: "the obliquity of the ecliptic must have remained sensibly constant throughout geological history"; and that the inequality caused by the weight of the great ice-sheets could not have altered the position of the Arctic Circle by even as much as half an inch. However, he did admit the possibility of polar wander within the body of the Earth. In the geologically short timescale of the ice ages, he considered that the poles could have wandered no more than 3°, but perhaps as much as 10° to 15° over the entire lifetime of the Earth.

On the subject of the Glacial Period, he mused that it may have been only *apparently* a period of great cold – if as proposed, the North Pole once stood somewhere near to where Greenland now lies. However, he still could not conceive of any geological process that could bring about such an outcome. In his conclusion, he seemed to rule out polar wandering entirely:

> If, then geologists are right in supposing that where the continents now stand they have always stood, would it not be necessary to give up any hypothesis which involved a very wide excursion of the poles?

After this eminent intervention the matter was seemingly settled. Subsequent geophysicists throughout the twentieth century merely built upon this base and little has emerged to contradict it. However, in 1931, another geophysicist highlighted a mistake in Darwin's calculations. Once corrected, even the displacements of the axis permitted by Darwin become too great to accept.[6] Geomagnetic evidence also confirms that the magnetic poles, when their positions are averaged over millions of years, have remained always close to the modern rotational poles. The axis of rotation, if it wanders at all, does so within a well from which it cannot escape.

Once the astronomical theories had gained favour, as an explanation for recent ice ages and as continental drift appeared to explain all the other geological anomalies, polar wandering became a mere curiosity. It was no longer needed. The conclusion that polar wandering is not possible has entered the textbooks as proven. In fact, the nineteenth century pioneers have left open a window large enough for us to climb-through. Geographic pole-shifts of the order of a degree or so are physically possible and even a change of axial-tilt is admissible – if we can find an astronomical agent powerful enough to administer the blow.

Day upon Day

If the concepts of polar wandering and axis-tilts are to be ruled out then how much more so, is the possibility that the length of day could have changed in recent millennia? We do not need a mathematician or a nineteenth century astronomer to point out the difficulties with this idea; common sense will suffice.

The total kinetic energy stored in a rotating planet is truly vast. In order to change the length of day from 360 days per year to 365 days per year requires the application of a force that is some 1.4% of the total kinetic energy.[7] Where indeed would we find such a "cannonball". If we allow that it would be moving at around the same speed as the Earth in its orbit then it would demand a comet greater than 100 miles in diameter, or an asteroid of similar mass. The energy of most recently observed comets, notwithstanding their capacity to inject dust into the atmosphere, could have only the most negligible effect upon the rotation.

Moreover, this describes only the perfect scenario, applicable if the force is applied along the equator. In order for the axial tilt to be affected at the same time, then some component must also have been expended along a meridian. This increases the total energy requirement still further.

If the energy is to be entirely transferred to the rotation then the projectile must strike at an acute or near-tangential angle. The mass requirement then further increases as the angle of incidence decreases. The required mass begins to exceed that of the largest comet ever observed, approaching that of some of the larger main belt asteroids. If something this big had struck us only a few thousand years ago then it should have triggered a mass-extinction event even more destructive than that which wiped out the dinosaurs. The recently discovered Chicxulub impact crater implies that a comet just 10–14 km in diameter was sufficient to cause a mass–extinction.[8]

A further possibility is that the projectile that struck the Earth could have been travelling at a much greater velocity. The kinetic energy increases directly with the mass of the projectile, but with the *square* of its velocity. Yet, here we run into another wall. The Earth moves in its orbit at a speed of 30 km/s. The so-called Near Earth asteroids, with Earth-crossing orbits, cannot much exceed this orbital velocity, although comets can. Even at the maximum possible, an unlikely head-on comet collision, the combined impact velocity could not exceed 72 km/s. A little calculation will show that even this most shattering of impacts would be hardly a drop in the great reservoir of energy that is the Earth's rotation. Yet, such an explosion would utterly devastate all life on the surface.

If we look to a projectile travelling even faster than this, then we encounter a further barrier. Any object travelling faster than 42 km/sec must be completely unbound. It would escape the Sun's gravity and cannot be a member of our solar system. In the entire history of observational astronomy, no comet has ever been observed to travel significantly faster than this limit. Should such a comet ever be observed then it must have formed around another star.

Should we then, as Darwin advised, give up the whole idea of changes to the Earth's axis and rotation?

If the reader has understood the arguments, and the pattern of evidence set out in the previous chapters, then the question simply will not go away. How will you resolve the coincidences that all suggest a pole shift around 3100 BC and a former year of 360 days? What is your alternative explanation? Single-subject academics may not recognise the signs from their limited perspective. They can simply shuffle their part of the problem into another discipline and ignore it. The catastrophist researcher, looking across many disciplinary boundaries, does not have this luxury. We must find an explanation.

Ptolemy and Hipparchus

The rotation of the Earth is affected by a number of external and internal factors of which the most widely understood is the *precession of the equinoxes*. Credit for discovering the precession is usually given to the

Greek astronomer Hipparchus of Rhodes in the second century BC, but whether it was known to earlier astronomers in India or Babylon, must remain largely a matter of conjecture. Precession manifests as an apparent westward drift of the stars by about one-degree in an average human lifetime.[9] As such, it is barely perceptible without reference to the accurate star charts of an earlier generation. Claudius Ptolemy, whose astronomy was to dominate western thinking for so long, seems to have done little more than update Hipparchus' star charts to allow for a further three hundred years of precession.

The cause of precession remained unexplained until Newton's physics supplied the answer. It is the gyroscopic effect of a gravitational torque, exerted by the sun, moon and planets upon the equatorial bulge of the rotating earth. This compels the axis to describe a cone about the pole of the ecliptic over a period of 26,920 years.[10] The torque is directly proportional to the obliquity of the ecliptic, which is currently about 23°; therefore, it may be seen that if the axis were not tilted then there would be no precession.

When the precession of the equinoxes was first discovered, the celestial equator crossed the plane of the Earth's orbit (the ecliptic plane) in the constellations Aries and Libra, which then marked respectively the spring and autumnal equinoxes. When the spring equinox was chosen as the origin of the co-ordinate system it was therefore termed 'the first point of Aries'; and this terminology remains in use although the vernal equinox has precessed into Pisces. Over the full 26,000-year cycle, the equinoxes will precess right round the zodiac. The precession of the equinoxes can be projected backwards in time. During the Egyptian Old Kingdom our modern pole star Polaris was some distance from the celestial pole, which then lay closer to the star a–Draconis; and still further back in Neolithic times, the spring equinox would have occurred in Taurus.

The eyes of the non-astronomer will probably glaze over at this discussion of precession and the desire to skip the entire chapter may be strong! The Earth's rotation is not a simple subject and cannot be made so. It is not easy to describe these astronomical and geophysical phenomena without recourse to the specialist terms and mathematics. However, this book is written for the generalist with an interest in prehistory and is not intended to be a pure astronomical or geophysical reference text. Terminology will therefore be kept to a minimum. A number of authors who have put forward various simplistic theories of 'lost civilisations' and alternative chronologies for ancient Egypt have drawn heavily upon the subject of precession to support their ideas.[11] Therefore, the student of prehistory must attempt to grapple with these effects in order to understand what is credible and what is not. Similarly geophysicists, if they ignore the ancient mythology, will fail to recognise valuable evidence about the physics of the Earth.

Over many cycles the precession of the equinoxes contributes to the variations of climate that we know as the *ice ages*; and over a period of 41,000 years the tilt of the rotational axis fluctuates by about 3° between 22° and 25° due to perturbation of the orbit. The variation in tilt means that the rate of precession must also vary.

In the eighteenth century, with the advance of telescopic astronomy, an English astronomer named James Bradley was the first to notice a small displacement of stars along the meridian. This oscillation is termed *nutation* and is never greater than 15". The pull of the Moon on the equatorial bulge gives rise to the nutation and the pole of the Moon's orbit moves in a circle about the pole of the ecliptic over a cycle of 18.6 years.[12] Also, as the plane of the Moon's orbit is slightly tilted with respect to the ecliptic, there are additional components with periods of half a year and half a lunar (sidereal) month.

These tiny motions of the rotation axis are all *forced* motions. In theory, they should persist indefinitely and are sometimes referred to as *secular* variations; they arise due to the constant gravitational attractions of the Sun, Moon and planets upon the Earth's equatorial bulge. The precession and nutation are measured relative to a reference frame that is considered fixed in space. Hereafter, I shall refer to movements of the rotation axis, relative to this inertial frame, as motions "in space".

The Earth's Free Wobble

In addition to the forced motions, the axis of rotation is subject to a number of free oscillations. These arise due to displacement of the axis of rotation from its rest position and its attempt to return to that stability. No external force is invoked; the non-specialist may like to visualise these by analogy with a pendulum that is displaced and allowed to swing freely, or a spring that is pulled and released. All such free motions have a natural period of oscillation and must ultimately decay away to their rest position. We usually encounter rotational oscillations in the spinning of a Rugby ball (or an American football) in flight, or in the nodding of a spinning top.

The free oscillation of the rotating Earth was studied in the eighteenth century by the Swiss mathematician Leonhard Euler who derived an equation for it. As a consequence of its rotation the Earth is a flattened oblate spheroid and may only rotate in a steady state about its polar axis.[13] Euler treated the Earth as a rigid body and concluded that it should oscillate with a natural frequency of about 305 days.

Throughout the nineteenth century, various astronomers (who would later come to be called geophysicists) looked for evidence of the Eulerian nutation – despite the fact that many geophysicists had already accepted that the Earth was not rigid, but best modelled as an elastic shell enclosing a molten core. Therefore when, in 1891, S.C. Chandler discovered a small variation of latitude with a period of approximately 14-months (428 days) it was soon recognised as the Eulerian wobble with its period lengthened due to the elastic yielding of the mantle. The 14-month nutation has henceforth been known as the *Chandler wobble*. Another oscillation with a period of exactly a year was explained by seasonal movements of the ocean and atmosphere. Both motions of the pole are very small indeed, of the order of 0.1" or about 30 metres on the ground.

Geophysicists measure the free oscillation as an apparent variation of latitude. A frame of reference is used which rotates with the Earth, carrying the observer around with it. The latitude of the observatory is measured from a north-south axis defined as fixed in the body of the Earth. Imagine that the axis of the rotating Earth is exactly aligned with the inertial frame in space then the declination of a zenith star (adjusted for flattening) would give the latitude of the observer. However, if the rotation axis wanders in space then there will be an apparent variation in latitude. Similarly, if the axis 'wobbles', or wanders within the body of the Earth then a similar discrepancy occurs. Henceforth I shall refer to movements of the axis within the body of the Earth as movements 'relative to geography' to distinguish them from those relative to the inertial frame. The loose term "pole shift" has a similar meaning. Some geophysicists also reserve the term nutation for the motions in space and use wobble for the body motion.

Some people may find it difficult to visualise the difference between these motions. Perhaps a good analogy might be to take a geographers' globe, with its usual polar mounting that shows the tilt of the Earth's axis; and give the globe a spin. The movements in space can be simulated by picking up the globe in its mounting and simply tilting the whole thing; the axis within the globe moves with it. To simulate the movements relative to geography, you would have to place the stand on a flat surface so that it cannot move; and then imagine that you could somehow move the spinning globe around within the mounting frame, such that the north and south poles are displaced.

Now the mathematics that describes the Earth's free oscillations has become very complex as geophysicists have included terms to allow for the elasticity of the shell (the crust and mantle) and the viscosity of the fluid core. Within the fluid metallic core, there is believed to be an inner core of solid iron and some models also allow for this. The theories are further complicated by the fact that many of the properties of the Earth's interior are not yet known with certainty, but have to be implied from these mathematical models.

It had been recognised since at least 1839 that this model of the rotating Earth as a solid shell enclosing a fluid core must give rise to two modes of free oscillation.[14] One mode is the Chandler wobble of the shell alone (in which the core takes no part); the other is a consequence of the fluid core that can only occur if the rotation axes of the shell and of the fluid core become misaligned; sometimes referred-to as the core mode. These theoretical modes of oscillation have been examined by various geophysicists from the late nineteenth century up to the present.[15]

The Chandler Wobble

To speak of the Earth's 'axis' is therefore a generalisation. The angular momentum of a rotating body may be expressed by a vector acting along the axis about which the body rotates at any instant; and which must maintain a fixed attitude in space unless disturbed by an external force; it is sometimes called the principal axis. However, free oscillations, such as the Chandler wobble, are caused solely by a redistribution of mass within the Earth and cannot alter the total angular momentum.

The stable or rest position of the Earth's rotation occurs when it turns about its axis of symmetry. For an oblate spheroid this is its minor axis, sometimes called the *figure axis* and the position of the figure axis is determined by the distribution of mass. Movements of the ocean, atmosphere, animals and of course earthquakes in the interior, all contribute to the position of the axis of figure within the body of the Earth. It may therefore be seen that the Chandler wobble can only be generated if a redistribution of mass drags the figure axis away from the axis of angular momentum. In geophysical texts, this may be referred to as an "excited" state. Since the Chandler wobble can be observed (albeit of small amplitude) then it must be continuously excited by rearrangements of mass on and within the Earth.

Once the axis of rotation deviates from the figure axis then the two must try to reunite and they hunt about the principal axis in an irregular spiral until they again coincide, or until another excitation occurs. The Chandler wobble should therefore be completely damped within a period of about 20-40 years.[16]

An important consequence of the Chandler wobble is that it causes *pole tides*. These arise because world sea level (the geoid or ellipsoid) must conform, with only a short time lag, to the instantaneous rotation; whereas the solid Earth remains symmetrical about the figure axis. The tiny modern Chandler wobble produces a variation of the sea of only about 3-4mm, which is virtually undetectable. Even so, some geophysicists believe that the friction of these pole tides in shallow seas may be enough to explain the damping of the Chandler wobble.[17] It may be seen that if the Chandler wobble were of only a little more significant amplitude, even a few minutes of arc, then these pole tides would exceed the height of the lunar tides and would inundate low-lying coastal regions. It was this phenomenon that some nineteenth century geologists, following theories of polar wandering, sought to employ as the mechanism for the Biblical Flood.

The Nearly-diurnal Free Wobble

For a long time the second theoretical mode of oscillation languished in the geophysical papers as a curiosity. However, if the Earth possesses one mode of oscillation then it must also possess the other. Because geophysicists were so familiar with their rotating reference frame, they tended to view it only in terms of its motion relative to geography. From that viewpoint, it is a retrograde circular motion of very small amplitude, so small indeed that until late in the twentieth century, no one could certainly claim to have detected it. The equations suggest that it should have a period of about 23 hours 52 minutes, hence it acquired the name *nearly diurnal wobble* – an unfortunate choice as it turned out. As the wobble can only occur if the principal axes of the core and of the outer shell have become misaligned, and since these are closely aligned on the Earth today, then one would expect the amplitude of the nearly-diurnal free wobble to be vanishingly small.

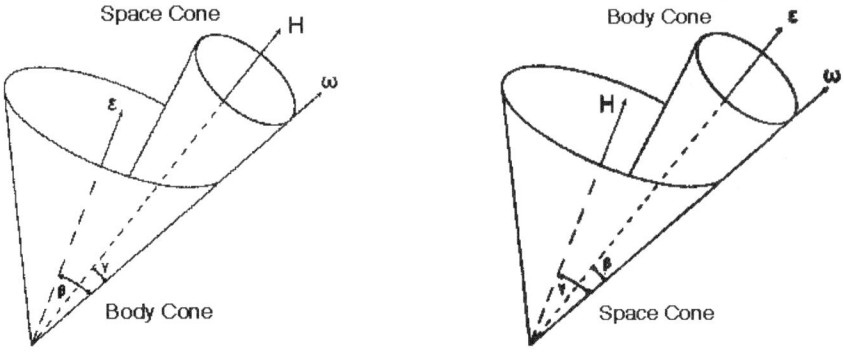

Figure 9.1 Poinsot Diagrams
Poinsot diagrams for the Chandler wobble and for the core (nearly-diurnal) wobble. While the Chandler motion takes place mainly in the body of the Earth, that of the core motion manifests mainly in space and so would not cause a significant pole tide.

The French mathematician Louis Poinsot devised a graphical method of representing the special and body motions of a rotating body. Applying this method to the Earth, the motion of the axis in space can be represented by a cone called the *herpolhode*, which traces the motion of the axis relative to the inertial frame. Similarly, the motion of the axis relative to geography can be represented by another cone called the *polhode*. This is a very useful diagrammatic representation that may help the non-mathematician to understand intuitively what is going on.

For the Chandler wobble, the body motion is much greater than the motion in space and so the Poinsot representation shows the space cone rolling along the inside of a much larger body cone. In fact this can be misleading; as noted above, the scale of the body motion is about 30m at the Earth's surface, whereas the cone in space amounts to just 3cm on the ground and has a period of exactly a day.[18] Therefore for all practical purposes we may condense the space cone to a line and represent the Chandler Wobble as a spiralling of the axis of rotation about the axis of figure.

However, for the nearly diurnal wobble the situation is reversed. Here it is the body motion that is small and the complementary motion in space that is much larger. The Poinsot representation shows a retrograde rolling of the body cone on the inside of the space cone. The theoretical period of the wobble is indeed nearly a day, but the nutation in space turns out to have a much longer theoretical period of roughly 460 days and should also be about 460 times greater in amplitude. The name *nearly diurnal* wobble is therefore scarcely appropriate for the motion as a whole.

Various investigators claimed to have observed the nearly diurnal free wobble with amplitudes of a few arc-seconds.[19] It was not until 1974 that another geophysicist, Alar Toomre, pointed out that none of these claims could be true.[20] They had all neglected that motion in space and any wobble of this magnitude would have to be accompanied by a nutation as large as that discovered by Bradley two centuries earlier, and which would surely be noticed by any competent amateur astronomer. This should be a lesson to us all that numerate scientists can be just as confused by their terminology as the rest of us; and that mathematicians do not always understand – and make little attempt to explain – precisely what their equations mean in simple language. Toomre suggested that *core nutation* would be a better name for the motion, but even this does not seem adequate to describe it.[21] In my own earlier research I called it *core-mantle precession.*

The nearly-diurnal wobble could be shown to be subject only to very weak damping by the Earth's magnetic field, which couples the core to the mantle. Once triggered, the core nutation should persist for a long time, something like 2,000-5,000 years. Compare this again with the Chandler wobble, which would die away in only 20-40 years.

Undeterred, the geophysicists went back to their data and looked for evidence of the core motion, this time in space, and discovered a nutation of amplitude 0.01 arc second with period about 434-444 days.[22] This is only some 30 cm on the ground, around a hundred times smaller than the Chandler wobble; and so the corresponding body wobble can only be some 2×10^{-5} arc second – which is a fraction of a millimetre. This is about as negligible as negligible can be! The shortening of the observed period from that predicted is analogous to the modification of the Euler/Chandler period and shows us that the mathematical models do not yet tell the whole story.

Certainly, today, the planet that we live on has a very stable rotation; and the uniformitarians therefore project backwards that it has scarcely changed over geological time, let alone during the human time frame. Although these nutations are of such small amplitude, they are fundamental rhythms of the Earth, which may become far more evident in the event that the world experiences an extreme geological catastrophe.

None of the various theories of polar wandering or mountain building could proceed without exciting these free nutations. If you would like to believe in the reality of the Great Flood, or the sinking of Atlantis, then these effects cannot be swept under the carpet. Survivors of a real cataclysm would experience the effect of these motions on the real Earth.

Earthquakes and Comet Impacts

There are various theories as to what excites the modern Chandler wobble. One factor may be the deformation caused by earthquakes. The Richter scale measures earthquake energies on a logarithmic scale from 0-9. The upper limit marks the point at which rock should disintegrate and it seems unlikely that quakes much above that of the Indian Ocean quake of 2004 (Richter 9) could ever occur.[23] Much research was performed following the severe quakes that took place in Chile during 1960 at Richter magnitude 8.7; and that of Richter 8.6 in Alaska in 1964, both along the tectonically unstable Pacific Rim. The Chilean earthquake resulted in the slippage of an immense block of land 300 miles long. The energies unleashed by these quakes were of the order of 10^{18} joules, equivalent to 31 megatons of exploding TNT.[24] It has also been shown that the poles shifted by 0.01" to 0.02" following these events.[25] We may also debate how much of this was due to the movements of the initial quakes and how much to secondary effects of the huge tsunami waves that these generated in the Pacific.

Now this result is interesting, for even the smallest of comet impacts, such as the Tunguska impact of 1908 have been estimated to release energy of equivalent magnitude. The Tunguska object exploded in the atmosphere and did not have an opportunity to cause seismic effects. It follows therefore that a slightly larger impactor reaching the ground must release a point source of energy *exceeding* that of even the largest terrestrial earthquakes and must also have the propensity to make the Earth wobble slightly on its axis.[26]

A deeper examination of the energy released by comet and asteroid impacts reveals some even more interesting figures. A study by Schultz and Gault into asteroid impacts at oblique angles (<45°) revealed that the most likely class of impactors, the so-called Apollo asteroids, could strike the surface at velocities of 12 to 15 km/s. As this is greater than the Earth's escape velocity (11.2 km/s) then perhaps 80% of the projectile would retain enough energy to bounce back into space, or to produce high velocity fragments that would orbit the Earth and reimpact over the next few days.[27] Another consequence of oblique trajectory is that the vertical component of the force is greatly reduced, resulting in a much smaller and elongated crater.

Near–Earth asteroids are constrained by orbital dynamics to velocities below 30km/s, but comets, which approach from the outer reaches of the solar system, can attain much higher velocities as they sweep by the Sun; furthermore, they may move in retrograde orbits and could therefore collide with the Earth head-on. In this eventuality relative impact velocities as high as 72 km/s are possible; for example the infamous comet P/Swift-Tuttle, which was described by Sky and Telescope magazine as "the single most dangerous object known to humankind" could one day collide with the Earth at a velocity of 60 km/s.[28]

In examining the period of human prehistory, we need not concern ourselves with full-on impacts by giant comets like P/Swift-Tuttle. If anything of this scale had occurred during human evolution then we simply would not be here to discuss the matter. Giant impacts occur only at intervals of many millions of years and mark the boundaries of the major geological epochs and major extinctions of species. The likely scale of impactors during the human period *must* have been small enough to have left no spectacular traces on the Earth today. The question is therefore whether these posses enough energy to have generated the kind of catastrophes that we find in our myths and legends.

One such example is the recently discovered Rio Cuarto crater field in Argentina. Here, a large meteor only 150-300 m diameter, of the type known as a *carbonaceous chondrite* impacted at an oblique angle, leaving an elongated main crater and a series of smaller ricochet impacts downrange over an area of 4 km. The angle of impact was certainly less than 15° in this instance and possibly less than 7° as it approached from the north-west.

However, the most interesting aspect of this impact is that even such a small meteor (for an assumed velocity of 25km/s) would release an energy thirty times greater than the Siberian explosion and therefore some thirty times greater than the Chilean earthquake. Moreover, the state of preservation of the craters and the condition of the recovered fragments all suggest an age of less than 10,000 years.[29] The world-wide effects of this event – whatever they may have been – were certainly experienced by our early-civilised ancestors. Moreover there are other scars that are also probably due to oblique comet or asteroid impacts, for example the Carolina Bays and the oriented lakes of northern Alaska, both of which have been loosely assigned a late glacial or Holocene date.[30] Since nearly three quarters of the world's surface is covered by ocean we may reasonably deduce a further 3-4 times as many strikes in the ocean during this period.

Studies of the long-term damage due to oblique impacts have mainly focused upon the disturbance to the climate and biosphere because of the mass of ejecta inserted into the atmosphere. These results come from experiments with small hypervelocity impacts in the laboratory, which are then scaled-up. From these studies, it may be seen that cometary fragments of around 2-3 km diameter, travelling at sub orbital velocities of 7-10 km/s, would release energies around 10^{21} joules.[31] Smaller bodies travelling at cometary velocities would release similar impact energies – but with a reduced capacity to release dust clouds into the atmosphere. These energies are of the order of a thousand to ten-thousand times greater than the largest terrestrial earthquakes. It may be seen therefore that these high velocity events have the potential to excite the Earth's wobble, but with conceivably much less dust insertion than a major volcanic eruption.

Consider the figures for a moment. A large earthquake (and its associated effects) can cause a pole shift of the order of 0.01-0.02 arc seconds. A small comet impact could therefore be expected to produce excitations of between 10^3 and 10^4 times larger, of the order of 10 arc seconds, up to 3 minutes of latitude. There remain of course many other variables. There is also the question of the composition of the impactor; how much energy is expended in crater formation; and how much is retained by the projectile as it ricochets into space.

Could even higher energy impacts have occurred during human prehistory? Of course, they could – but then one would have to face the question of where is the crater? Giant comets of the scale of the K/T-impactor enter the inner solar system roughly every 200 years.[32] Perhaps the best example would be comet Sarabat of 1729, which may have had a nucleus of 100-300 km diameter, or the enigmatic Chiron, which orbits between Saturn and Jupiter.[33]

Until very recently, hardly any research was done on the characteristics of an impact at sea. Like so many other aspects of this problem, it was a no-go area for respectable academics until the Jupiter impact and the discovery of the Chicxulub crater. It seems the asteroid that killed the dinosaurs at the end of the Cretaceous epoch fell, at least partly, into a shallow sea and produced a tsunami some 50-100 m in height.[34]

Once again, this phenomenon is not restricted to large impacts. The 1960 Chilean earthquake produced waves that swept right across the Pacific Ocean and caused loss of life in both Hawaii and Japan. Another recent study has highlighted that an impact by an asteroid only 400m in diameter, falling into deep ocean water, would generate a tsunami with a run-up of some 60m when it reached shallow coastal waters. If such an impactor were to strike the mid-Atlantic then these waves would simultaneously reach the North American and European coasts, generating tsunami that would sweep inland submerging low-lying coastal countries such as The Netherlands and Denmark.[35] A 100-m tsunami would penetrate 22-km; while a 200-m wave could travel some 55-km inland.

These dangers scarcely need to be overstated. The authors conclude with the statement: "The entire research field of geologic assessments of tsunami produced by impactors is virtually non-existent and needs to be initiated". This must surely rank among the greatest understatements ever to appear in a scientific paper!

Pole Tides – a Model for the Great Flood?

A tsunami is all over in a day and its victims will not care what happens next. If the same impact event also causes the Earth to wobble on its axis then it must also generate a pole tide. These have the potential to dwarf even the destructive power of a tsunami and would have a far longer lifetime.

As was argued above, an impact by a relatively small cometary body, allowing for all the uncertainties, could generate a wobble of a third, to perhaps as much as a whole degree of latitude, although smaller shifts are far more likely. What does this mean? It means that the pole of figure would become separated from the pole of rotation by that amount and the sea level world-wide must rapidly conform.

The pole tide generated by such an event would be equivalent to the change in the Earth's radius for whatever latitude is considered. Moreover, the magnitude of sea level change must follow a pattern of alternate quarter-spheres. In the direction that the pole has shifted the shoreline in that quarter-sphere would be lowered; in the quarter sphere behind it,

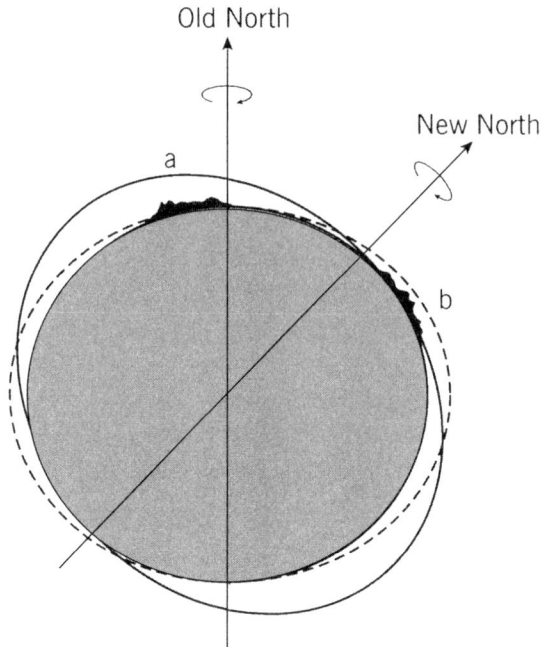

Figure 9.2 Pole Tides
An illustration of pole tides caused by the Chandler wobble as a model for the Flood of Noah.
The sea level is determined by the rotation axis, while that of the land is determined by the
figure axis. The tidal bulges sweep round the world with the 14-month period of the wobble.
However, if the solid-earth can deform more quickly than this (and since the wobble decays
exponentially) then the Flood would only devastate two opposite quarter-spheres of the world.

the sea level would rise by a similar amount. Each effect would be repeated at its antipodes, giving two opposite quarter-spheres where the sea level rises and two where it falls. The greatest effect would be along the meridian of movement, but at 90° east and west there would be a neutral meridian, where the effect is zero. These effects would then sweep round the Earth with a period of roughly thirteen-fourteen months, decreasing exponentially in amplitude.

The likely height of these pole tides can be roughly estimated. The difference between the equatorial and polar radius is 21,383-km. This flattening depends, of course, upon the speed of rotation, which itself may be altered by such an event. The difference for a degree at mid-latitudes, assuming no change of flattening, is therefore *of the order of*

373-m; similarly a shift of half a degree represents 186-m; a tenth of a degree represents 37-m and so on. These estimates are of course simplistic, as there remain so many other variables; the thin crust of the ocean floor is also pliable and could be expected to adjust its figure in a series of massive quakes. Therefore, one may visualise how some areas of the coastline might be deeply submerged, only to be elevated again within a few years, or even months, as the wobble decayed. Elsewhere the opposite phenomenon would occur. Land would be raised out of the sea, only to sink back again. Consequently, the permanent position of the figure axis must reconfigure and after some 20 years, the poles of rotation would settle to these new positions. Thus, the Chandler wobble resultant from an impact event can, and indeed must, bring about a permanent pole shift.

What then, of the axis of rotation in space: the obliquity of the ecliptic? The Chandler wobble does not affect the total angular momentum, which remains unchanged in the absence of external torques. The spatial motion associated with the Chandler wobble remains no greater than a few centimetres on the ground. However, if the Earth's rotation were caused to wobble in this mode then it must also wobble in the core mode, but how could the axes of the core and the crust become misaligned?

Specialist studies consider only the tiny core nutation of the present day. One explanation suggested for this has been magnetic torques generated in the fluid core, which act upon the solid inner core. Once misaligned, the core nutation must occur. Another possibility is that quakes deep within the mantle could alter the figure of the core-mantle boundary enclosing the fluid core, thus exciting the wobble.[36] However, there is no reason to believe that the magnitude of these effects would ever be more than a curiosity for life on the surface.

An impact on the surface is itself an external torque: an impulse. However, unlike the gravitational torques of the Sun and Moon, which attract the whole planet, the impact acts only upon the outer shell. It has the capacity to rotate the crust and mantle against the fluid core; and for a high-energy impact this could become significant, particularly if the impact trajectory is oblique. In this case the angular momenta of the core and outer shell would become misaligned by some small angle and would begin to precess about each other. These are the conditions for the core nutation or nearly-diurnal wobble as discussed above, which would then manifest; and at the end of this episode the tilt of the ecliptic would be permanently modified. It is important not to overstate this however: unlike the case of a geographical shift, any *significant* change to the Earth's angular momentum demands a very high-energy impact. Even a head-on collision with an average comet would not posses enough energy to alter the obliquity by more than a tiny fraction of a degree. A tilt of about $1°$ requires an impact energy some 10^9 times greater than the largest known earthquake.

It may be seen that this combination of external torques due to oblique impacts, together with the presence of a fluid core within the Earth challenges the long-standing uniformitarian assumption that the obliquity of the rotation axis has scarcely changed since the Earth's formation. It becomes instead a question of how many such impacts could there have been and what was the magnitude of change in each instance? In considering the shorter period of human prehistory, we may set aside the statistical probabilities. A once-in-a-million-years event could occur twice in the same day – that is in the nature of probabilities. We need only look back at the record of human prehistory: physical evidence as well as history and myths, to see if we can recognise the pattern of evidence.

The Failed Theories

If the pole shift hypothesis founders upon the rock that is the kinetic energy of the Earth, then perhaps there is some other way to deliver the energy, or to reduce the energy requirement.

Such a concept was crustal slippage, first proposed in the nineteenth century and revived by Charles Hapgood in the late 1950's. He proposed that parts of the Earth's lithosphere could become out of balance due to geological movements.[37] Perhaps the principle of isostasy did not always apply and that when the strains became too great, the entire crust (or parts of it) could slip against the mantle.

The advantage of Hapgood's approach is that only sufficient energy is required to move the crust, not the entire planet. It is supplied in the form of potential energy, stored up by gradual processes and then somehow, suddenly released. Unfortunately, this will not work. The theory is just problem transference again and the causes of the initial imbalances still need to be explained.

Imbalances of mass within the interior could not build-up because they must excite the Earth's wobble. Geophysicists of the mid-twentieth century understood this phenomenon. Such 'imbalances' would be dissipated as tiny shifts of the pole via the mechanism of the Chandler wobble, or as adjustment of the crust at the plate boundaries by ordinary earthquakes. Indeed this is what earthquakes are. Earthquakes excite the Chandler wobble.

Physicist Peter Warlow looked even deeper for a mechanism and found it in the molten nickel-iron core. At a depth of 2900 km lies a transition layer where the silicate mantle ends and the nickel-iron core begins. Deeper down, at a depth of 5150 km is a solid core, which may be pure iron. Seismologists infer the existence of the fluid core because the shock waves produced by earthquakes cannot penetrate it.

Warlow proposed that the molten core acts as a lubricating layer allowing the entire mantle to slip against the core.[38] Indeed, he believed that at times the outer shell might completely turn-over, such that the south pole could replace the north pole; and that this was the mechanism for the periodic reversals of the geomagnetic field that geologists find in the record of the rocks.

One advantage of this theory is that the energy requirement is reduced to that sufficient to move the crust and mantle alone, which makes up only about two-thirds of the Earth's mass. Warlow supplied this force in the form, not of a cometary collision, but a gravitational interaction with a rogue planetoid, inevitably inspired by Velikovsky's rogue 'Venus'. Unfortunately, this mechanism will not work either. Tidal interaction with a passing planet should be of only the same order as that which the Moon exerts. Gravity would shake not only the container (the crust and mantle) but would also act upon the deep core.

The detection of the fluid core also allows us to reject the theory of crustal slippage. In order for the crust to slip over the mantle, it would require a lubricating layer between crust and mantle. This should have the same effect upon seismic waves as does the fluid outer-core; and seismologists affirm that no such layer exists. The modern theory of plate tectonics shows that the continents do indeed move around, but this has occurred gradually over millions of years, not rapidly over days and weeks. Crustal slippage is a theory that sounds quite promising, so long as the arguments are kept suitably vague.

The New Planet

In 1781, the British astronomer Sir William Herschel chanced upon the first new planet to be discovered since ancient times. He thought he had found a comet!

As astronomers turned their telescopes to the new planet Uranus, they observed that it rotated at an angle only 8° from the plane of its orbit – in fact, its axis is tilted at 98° and it rotates nominally retrograde. As Uranus progresses in its orbit, we see first one pole, then forty-two years later, the opposite pole faces us. It is an extreme parody of our own axial tilt.

We may be sure that whatever happened to leave Uranus rotating on its side, happened during the early history of the solar system, because all of its moons rotate in the same plane. Yet, Uranus cannot have formed this way from the solar nebula. Something must have knocked it over and its satellites have reformed according to the changed dynamics.

Uranus is a gas giant some fourteen and a half times more massive than the Earth. The best explanation for the tilt is that shortly after its formation Uranus was struck near one of its poles by a planet with about 7% of its present mass – nearly as large as the Earth.[39] It may be that the present satellites aggregated from the disrupted remains of this hypothetical planet.

An impact near one of the poles (or with a strong component along a meridian) is the first requirement to cause a change in axial tilt.[40] The gyroscopic principle tells us that a rotating body is stable about its principal axis of rotation and that a large and constant force must be applied perpendicular to the axis to make it precess. When the force is removed, the precession ceases. In an impact collision, this force is applied as an impulse of very short duration, which causes an instantaneous jump in the tilt of the axis.

While astronomers and physicists may be reticent to consider a recent catastrophe on the Earth, they may safely discuss such matters during its formation, or on other planets. Thus, the impact events and their effects are kept safely distant in time and space from human experience. This body of research is also available to the catastrophist researcher, for the physics applies just the same.

Astronomer Sir Fred Hoyle also saw an anomaly in the rotation of giant Saturn, which is the most rotationally-flattened planet in the Solar System.[41] It too, must have formed from at least two coalescing planets of substantial mass, which left it with an axial tilt of 29°. Its magnificent ring system perhaps tells us that this encounter may not have been so ancient, in geological terms, as is often assumed.

Another example is the anomalous rotation of our twin planet, Venus. Its dense atmosphere kept its length of day a secret until radar imaging by space probes showed it to be rotating slowly retrograde, with a period equal to 243 of our days – longer in fact than its year! Venus too, must have suffered a major collision, in this case against the spin, which completely stopped it and indeed sent it into reverse. However, as Venus has no satellites there is currently no proof that this event was ancient.

Closer to home, astronomers now believe that our own Moon was formed by a collision between a Mars-sized planet and the proto-Earth – the so-called "giant-impact" theory.[42] The Earth-Moon system then reformed, as we know it today. We may infer that the Earth's axis has been further tilted since the formation of the Moon, as its orbit is now inclined to the Earth's equator.

Therefore, we need not doubt that our planet could adjust its figure rapidly to a change in the rotational characteristics of the order of five days per year. After the formation of the Moon, the rotation was much faster and estimates based on fossil corals show that during the Devonian era there were some 30.7 days in a month, corresponding to about 400 days per year.[43] Tidal retardation, as was discussed in the chapter on eclipses, tells us that the plate structure of the crust is perfectly able to cope with a much faster rotation. In the event of a collision, any accumulation of ocean water at either poles or equator could be only temporary as the 'solid' crust would conform after only a short interval.

The Post-Chicxulub World

Now that we have seen a comet strike mighty Jupiter and, at last, we have proof that an impact event caused the great Cretaceous extinction, many doors have suddenly opened. The obvious has been proved. Comets and asteroids can hit the Earth! In this new environment, a number of recent books and studies have provided valuable input.

Astronomers Victor Clube and William Napier made their case as far back as 1982 that the periodic comet Encke was the prime suspect for recent assaults against the Earth.[44]

Encke's Comet has the shortest period, at 3.3 years, of any periodic comet. It was actually seen several times before Johan Encke determined its orbit in 1822. Having passed-by the Sun so many times it is never spectacularly bright; and there is currently no accepted theory as to how it reached its present orbit. Since its aphelion lies well within the orbit of Jupiter, it cannot have been perturbed out of a long period orbit by an encounter with that planet. Neither can it have been in its present orbit for very long, or it would have lost its icy tail. Current estimates suggest that it assumed its present orbit no more than five thousand years ago.[45]

According to Clube and Napier, the progenitor of Comet Encke would have been a spectacular sight around 3000 BC. They, and other astronomers, argue that it is but one of a swarm of bodies that are now spread out along its former orbital path. Although the modern Encke, with a mass of some 10^{13} kg can be at most 1-2 km in diameter, its progenitor may well have been a huge object up to 100 km across.[46]

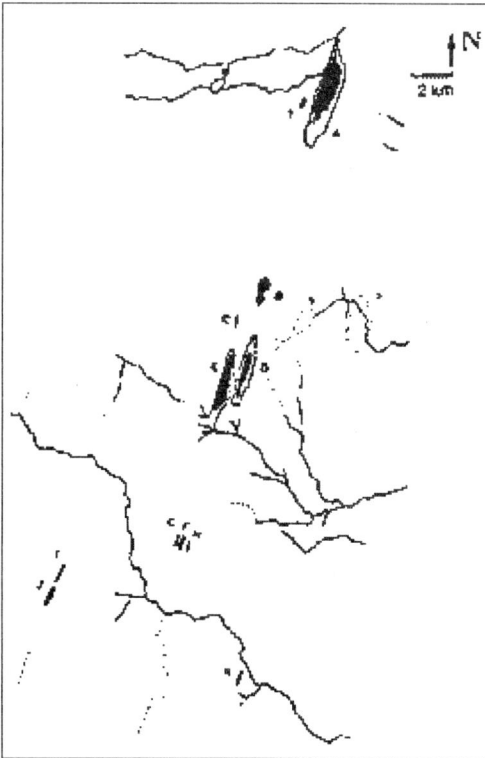

Figure 9.3 A diagram of the Rio Cuarto crater field in Argentina
(Reproduced from Schulz and Lianza, Nature 355, 1992, p. 235)

Indeed, with such a short orbit and regular bright returns, Comet Encke should have entered mythology as at least an equal of the planets. This takes us back to the origins of Indian and Babylonian astrology. The Babylonian sky-gods Anu and Enlil are often depicted along with the visible planets and it may be that they represent cometary bodies that were temporarily prominent.[47] As the progenitor broke apart, the comets faded and the origins of Anu and Enlil were forgotten.

Clube and Napier would also argue that fragments of progenitor-Encke may have struck the Earth during Neolithic times, giving rise to much of the fear of comets that has been handed down via the classical writers.

In his recent book *Impact!*, Gerrit L. Verschuur picked up on this theme and linked it to the climate changes that occurred around 3000 BC. If one may quote him:

> In later times, the nature of the sky gods began to change [*there follows a quotation from Clube and Napier*] ... because the skies themselves cleared of the dangers that had seemed so obvious to those who lived 5,000 years ago. In due course, the notion that such dangers ever existed was expunged from astrology, which was then left with the system of belief about the role of astronomical phenomena in guiding human destiny that made no sense any more. Astrology had to redefine itself, which led to its inevitable decline into the useless set of beliefs peddled today in the guise of horoscopes.[48]

Yet despite his enlightened stance, even this modern author remains committed to the view that it was the insertion of cometary dust into the atmosphere, which led to the demise of the ancient world. In a discussion of nineteenth century theories about the likelihood of comet impacts, he cites the works of Laplace and Milner. The Reverend Thomas Milner, in 1860, wrote a long treatise on the prevailing theory that a cometary encounter may have caused the Deluge. He was unimpressed by Laplace's view of an axis-tilt, believing comets to be wholly gaseous and ethereal; and that the Earth would pass right through a comet with no effect.

Commenting upon these two extreme views, Verschuur neatly summarises the present state of opinion regarding asteroid and comet impacts:

> What Milner failed to appreciate was that the crucial aspect of cometary impact is not the ability of the earth's solid mass to shrug off the effect of the blow, but the inability of the atmosphere to absorb the energy of collision. A cometary impact will cause only the merest flicker in the earth's orbit, but its effect on the atmosphere is something else entirely. The atmosphere cannot absorb the energy without substantial changes to its physical and chemical makeup. That is where the danger to civilization of comet impact lurks.[49]

Yet, the modern catastrophist should not be satisfied even with this enlightened conclusion. The Chicxulub impact was so damaging because it was a collision by a massive asteroid at an encounter speed close to the Earth's escape velocity. A low velocity encounter would indeed result in most of the extraterrestrial matter, as well as the ejecta from the crater, being deposited back into the atmosphere. However, a more energetic strike would instead result in most of the impacted material being carried into space. We have the perverse situation that a low-energy impact might actually be *more* likely to cause a mass extinction than a high-energy impact. Yet, a high-energy event is required to disturb the Earth's rotation.

There is a further anomaly. Not only are the sky gods recalled in ancient mythologies, but we also find aspects of the Earth's fundamental rhythms preserved in the most ancient calendars. One may dismiss the ancient belief in a 360-day year as bad astronomy, but how do we explain the multiples of 432 that we find in Indian and Babylonian astronomy? The interval of 432 days is very close to the natural period that modern geophysicists calculate for the Earth's wobble – yet the motion is barely observable even with modern instrumentation. We find seven-year cycles in the climate, such as those of the Biblical Joseph story and the Gilgamesh Epic, which correspond to how the world really would behave if a wobble of the axis were superimposed upon the seasonal variations. How did ancient people learn of such things unless they constitute a memory of real experience?

References to Chapter Nine

1   Airey, G.B., "Change of Climate", *Athenaeum* 22.9.1860, pp 384-5.

2   Evans, J. Quart. Journ. Geol. Soc. 1876, xxxii, proc., p 108.

3   Twisden, Quart, Journ, Geol. Soc, 133, pp 40-41.

4   This alludes to the pole tide that would be produced by the Eulerian rigid-body wobble of period 10 months. The true fourteen-month period would not be discovered until 1892.

5   Darwin, G.H. (1877) Philosophical Transactions of the Royal Society, Part I, Vol 167, pp.271-312.

6   See Jeffreys, Sir H. *The Earth*, sixth edition, 1976, pp 478-481

7   Professor Hoyle calculated that the Earth's angular momentum would require prograde collisions involving only some 2% of the Earth's total mass. See: *The Cosmology of the Solar System* (1978) p 71.

8   Morgan et al, 1997, Size and morphology of the Chicxulub impact crater, *Nature*, 390, 4 December 1997, p 476.

9   Currently 50".27 per year, or about 1° every 72 years.

10 Strictly, this is called *luni-solar precession*. The much smaller torque exerted by the other planets act to change the obliquity of the ecliptic and is termed *planetary precession*; the two together being termed *general precession*.

11 For example, Graham Hancock in *Fingerprints of the Gods* would cite the alignment of various ancient monuments as an indicator that they preserve a memory of the sky, as it was 10,500 years ago; and Robert Bauval more credibly dates the Great Pyramid by the alignment of the 'air shafts' with the position of stars during the Old Kingdom era. Precession must also be taken into account when considering the astronomical alignment of British and Irish Neolithic monuments.

12 This is the same motion that causes the major and minor standstills of the Moon as discussed with references to Stonehenge and other Neolithic alignments.

13 I have examined this problem in more detail in chapter 5 of *The Atlantis Researches*.

14 Hopkins (1838)

15 See note 19 below (Rochester et al).

16 Lambeck (1988) p552

17 Lambeck (1988) p559

18 Vicente (1980) p 140

19 A list of these is given in Rochester et al (1974) p 355 and a historical note in appendix A, p 361 of that paper.

20 Toomre (1974)

21 Various attempts have been made to tighten up the terminology but they have not served to reduce the complexity, for example see the discussion in Melchior (1980). In my own earlier work, finding no suitable name for the nutation in space in the specialist papers, I described it as *core-mantle precession*, to highlight that it affects the mantle as well as the core. Core-mantle nutation might have been a better choice, as strictly, precession implies reaction to a force. The core motion discussed here is a free nutation, albeit one with a long lifetime and perhaps it is better just to record that these terms are all views of the same motion.

22 Capitaine & Xiao (1981)

23 As this book goes to press, it has been estimated that the quake and tsunami of December 26 2004 has excited the Earth's wobble and may have relocated the island of Sumatra southwest by as much as 20 metres.

24 Jeffreys (1976) p136

25 Lambeck (1980) table 2 and Lambeck (1988) pp565-6

26 See the discussion in chapter 13 of Hoyle (1978), The Cosmogony of the Solar System.

27 Schulz & Gault (1990)

28 Schaaf, F (1997) p 154

29 Schulz and Lianza (1992)

30 See, for example. Prouty (1952)

31 Schulz & Gault (1990), p255

32 Levy 1995, p207, citing Shoemaker

33 Bailey, Clube & Napier 1990, p435-6

34 Bourgois et al (1988)

35 Hills et al (1994)

36 Smith, M.L. (1977) p138

37 Hapgood, 1958, *The Path of the Pole*, p 41.

38 Warlow, P. (1982), *The Reversing Earth*.

39 Greenberg, R. 1975, *Icarus*, 24, pp325-322

40 For an easy to understand mathematical expression of this, see Mary Lunn, A first course in Mechanics, Oxford Science Publications, 1991, pp 170-1.

41 Hoyle, F., *The Cosmology of the Solar System* (1978) p 66..

42 Haartman, W.K., Phillips, T.T.A. & Taylor, G.T. (1986), *The Origin of the Moon*, Lunar and Planetary Institute, Houston.

43 For a summary see Lambeck, K., (1988), pp 610-614.

44 Clube, V. & Napier, W. *The Cosmic Serpent*, Faber & Faber, London (1982).

45 Schaaf, F. 1997, p 138.

46 Bailey, M.E., Clube, S.V.M., & Napier, W.M, The Origin of Comets, Pergamon Press, Oxford, 1990, pp 435-6.

47 Ibid, p 16-17.

48 Verschuur, G.L. (1996) Impact!: The Threat of Comets and Asteroids, Oxford Univ. Press, p 101.

49 Ibid, p 82.

10

Comets, Impacts and the Unstable Earth

The preceding chapters have demonstrated that, in ancient times, our pleasant stable world was disturbed by a catastrophic event. The memory of that worldwide dislocation is now so degraded that it has become shrouded in myths and religious imagery. We have observed that many clues consistently suggest a shifting of the rotational poles and a change to the length of day around five thousand years ago. Academics of various disciplines will unite to say that this cannot happen. Myths and legends, they will say, are not evidence. They would argue that comet or asteroid impacts are not powerful enough to cause such a change. They will point out that there is no recent crater and no signature of a recent giant impact; if it had happened then surely we would all be extinct. Therefore, it has not happened.

The uniformitarian view prevails. For scientists, the causes and effects that we see in the world around us are the only ones that there can ever have been; and so they may be projected as far back into the past as hard evidence allows. Where such evidence ends, we may not speculate. A theory of catastrophism in human prehistory is therefore not complete without a mechanism that could cause it. As usual, before we can suggest such causes we must first examine the orthodox view of Earth science and show that the proposed mechanism does not violate the known laws of physics.

The Signature of a Catastrophe

There has been much recent discussion of the K/T-impact that occurred at the Cretaceous-Tertiary transition, some sixty-five million years ago. Studies of the Chicxulub crater conclude that the impactor was probably a comet with a heavy metallic core. The impact after all, was first theorised a decade before the discovery of its crater, when geologists detected the presence of an iridium-rich deposit at the boundary layer.[1] This heavy metal is rare at the Earth's surface but more common in meteorites. Clearly no similar impact can have occurred during recent millennia or the iridium signature would be everywhere. We therefore have to consider circumstances in which an impactor could impart a high energy impulse to the Earth without suffering complete disruption in a crater-forming impact.

While many comets may contain heavy metallic cores, there is no reason to suppose that they all do. Recent encounters may therefore have been with bodies composed mainly of dust, ice and other frozen volatiles that would not leave the expected iridium signature. However, it still seems unlikely that these possess sufficient mass. The most likely scenario is therefore that the impactors were fast comets that encountered the Earth at a shallow or near tangential angle. The impact would then ablate only the outer layers of the comet, with its rocky core being carried away into space as it rebounded, probably as a number of hypervelocity fragments. This outcome is quite unlike the K/T impactor, which seems to have deposited most of its mass in the atmosphere as dust and ash.

Furthermore, the absence of a recent large and obvious scar on land demands that we should seek evidence of recent impacts on the ocean floor. Again, we can safely say that these must have been shallow angle impactors that rebounded into space. In 1981, an iridium anomaly was discovered in the southern Pacific Ocean and identified as a 2.15 million-year-old asteroid scar that may be associated with the close of the Pliocene epoch.[2] Named the Eltanin asteroid, after the survey ship that discovered it, this remains the only hard evidence of such an impact in the ocean basins. It is possible that other, more recent, anomalies remain to be discovered. A vast area of the southern oceans, most of the Pacific, Atlantic, Indian Ocean and even the Arctic coasts were devoid of human occupation until the last few millennia. Any impacts in these regions would have escaped mention in human myths and legends, although it is quite likely that their indirect effects may be manifested in the climate and sea level record.

If a comet did indeed strike the ocean during human prehistory, at some remote location, then the communities closest to it would have suffered worst from the immediate effects of ejecta and tsunami.

Conceivably, no observer in the vicinity lived to record the event. Further away, where more communities survived, the indirect effects may have been associated with the acts of a malevolent sky god.

The most devastating worldwide effect would be the pole tides associated with the Chandler wobble. These would persist for some twenty years as the rotational pole spiralled about the pole of figure. The worst inundations of the sea would occur in the first few years, returning gradually to normal as the wobble decayed. Away from the coasts, the combination of the annual seasons with the wobble would also produce unseasonable climate effects varying with a seven-year rhythm.[3] This would affect the growth of vegetation and crops, ice caps and glaciers; also river and lake levels. Only at the end of this transitional phase would the permanent changes become clear.

A wobble should also trigger massive quakes in the Earth's interior, together with world-wide vulcanism, which would further compound the climate disturbances. Both should persist beyond the initial twenty-year wobble. Consequently the figure axis would migrate and the rotational poles would ultimately settle to a new position on the surface of the Earth.

In parallel with the Chandler wobble would be the nearly diurnal wobble or 'core nutation'. We may safely ignore its tiny contribution to the geographical position of the poles and its associated pole tide. Instead, we should consider its much more significant spatial motion. Again, any variation in tilt should manifest as seasonal effects and we should therefore expect to see these as long term trends in the ancient climate record.

However, it is the lifetime of the phenomenon that is significant here. Current theories suggest that the variation of the tilt would decay only after a period of several centuries, perhaps 2,000-5,000 years. Over such a long period human civilisation can recover. We should therefore expect some evidence of any such motion to be preserved in climate data; and in ancient astronomy and calendars, as has been argued in the previous chapters. The irregular spiralling of the celestial pole would be very obvious at first, but becoming less and less noticeable over generations. The period, according to modern observations of the nearly-diurnal wobble, should of the order of 440-460 days – but this figure is derived solely from mathematical models. The evidence of the Indian calendar and the Babylonian sources suggest that it was actually experienced as a 432-day rhythm.

The earliest classical astronomy of the first millennium BC, as recorded in the star charts of Hipparchus and Ptolemy, does not offer any clear evidence for such a motion of the celestial pole. There can be no doubt that the rotation was stable by the time that Hipparchus discovered the precession of the equinoxes. The evidence of ancient eclipses takes us back a little further than this, to around 500 BC or perhaps even back to the Pyramid age. We may therefore safely conclude that any *significant* impact catastrophe must have occurred *at least* two thousand years before Hipparchus, at around 2500-3500BC, in order for the nutation to have completely decayed by the classical era.

However, even impacts of the magnitude discussed so far are simply not energetic enough to cause a significant permanent change in the obliquity – the tilt of the axis in space, still less to the length of the day. The core motion is only a transient phenomenon; whatever its amplitude, even if it persists for a few thousand years, the rotation must ultimately settle back very close to the axial tilt that existed before the impact. Yet, the episodes of climate change since the mid-Holocene and at the close of the ice age argue for a *substantial* non-secular change in obliquity.

Most previous theories of pole shift have foundered upon this rock. To gain a sense of this, consider the energy of a comet compared to the vast kinetic energy of the Earth's rotation. Even at the maximum velocity of 72 km/s, the impactor would have to be as big as the largest of observed comets and this lies beyond the threshold of major geological extinctions. If one of these struck us even a mere glancing blow in the ocean, it is hard to refute that it should have left some physical evidence.

Are impact velocities greater than 72km/sec attainable? The answer must be: *yes they are* – but they become increasingly unlikely! A few comets have been observed with velocities slightly higher than parabolic. These are termed *hyperbolic* orbits, implying that they are unbound and must ultimately escape the Sun's gravity. Since comets are observed to be ejected from the solar system by the perturbation of planets, then from time-to-time the solar system must encounter comets that originated around another star and have been similarly ejected. Since the Sun orbits the galaxy with a velocity of 20km/s then 'foreign' comets with at least this much excess velocity are theoretically possible – giving head-on impact energies up to say 90-100 km/s.

There is just one problem with this hypothesis. No comet has yet been observed to possess a *significant* hyperbolic velocity. The small excesses that have been observed can all be satisfactorily explained by planetary perturbation of long period comets. Neither have any meteorites been observed to possess these galactic speeds. Another remote possibility is that one of the planets (most probably Jupiter) could hurl a long period comet towards us with a hyperbolic velocity.[4] This is analogous to the 'gravity assist' that has been deliberately employed for space probes such as the Pioneer and Voyager missions to the outer planets. Any such comet would, by definition, be travelling close to the plane of the ecliptic and therefore have the potential to strike us tangentially near the poles.

The view of professional astronomers must in essence be uniformitarian; that is to say, they extrapolate the conditions prevalent in the modern solar system and calculate the probability of events in the past, based on the realities of the present era.[5] No one can be certain that the flux of comets was the same before the advent of renaissance telescopic astronomy less than 400 years ago. There may be many cosmic phenomena that operate on far longer cycles. It may be that the population of interstellar comets is not homogeneous and the Sun may currently be passing through an abnormally empty patch of space.

All uniformitarian assumptions are only as sound as the base from which they are extrapolated. Until 1994, the likelihood that we might observe a comet striking the surface of Jupiter would also have been considered unduly speculative. If a bright comet on a hyperbolic trajectory lights up our skies in the next few years, will this have a similar galvanising effect upon the academic community?

Galactic Supernovae

Astronomers accept that all the matter that comprises our solar system, certainly all the heavy elements, was created in the explosive destruction of ancient giant stars. Several billion years ago part of such a nebula collapsed under its own gravity to form the Sun, the solar system and ourselves. Astronomers like to discuss the formation of the solar system and the Earth as if it were something that ended billions of years ago. Nothing is permitted to be geologically recent. In reality, the process of accretion continues with every comet that strikes one of the planets – right up to Shoemaker–Levy 9 in 1994.

Supernova explosions in the solar neighbourhood are rare events. Only six have occurred within recorded history but astronomers routinely study them in distant galaxies. When a giant star reaches the end of its life, it begins nuclear fusion of heavy elements in its core, forming iron and all the heavier elements. Ultimately, when all this fuel is exhausted there is nothing to prevent the star collapsing. In an instant its core collapses to an incredible density and the in-falling material rebounds into space in the huge explosion that we call a supernova[6]

Several nearby supernova remnants are known. The ejected matter can be seen to expand away from the collapsed star with velocities as high as 10,000 km/s, or even 20,000km/s.[7] Most of this material is just hot dust and gas that expands until it collides with the sparse interstellar matter, which eventually halts the expansion. After several thousand years, the nebula becomes too diffuse to be lit from within and fades from our view. The largest known remnant is the so-called Cygnus Loop, which, on photographic time exposures, subtends an area of the sky greater than the full moon. From its rate of expansion, astronomers estimate that its parent star exploded some 40,000 years ago.

Although not often discussed, an expanding supernova remnant *must* also contain millions of larger bodies along with all the gas and dust. There is good reason to believe that supernova ejecta could contain bodies that we would recognise as meteorites and comets. The disrupted star must have been surrounded by its own cloud of comets, asteroids and meteorites, similarly disrupted and blasted into space at supernova velocities. Once ejected, these 'comets' should proceed to infinity at their expelled velocity. Even the star's planets, if it had any, once deprived of the central source of gravity, would depart tangentially into interstellar space at their former orbital velocities.

While the interstellar medium may halt the expansion of a gas cloud, clearly, nothing but the gravitational attraction of another star is going to hold back the asteroid and comet-sized bodies in an expanding supernova remnant. The solar asteroids have been orbiting through the solar wind for billions of years and they have not been brought to a halt. It is reasonable to suggest that the smaller and medium-sized bodies in an expanding supernova wreck might attain velocities of 8,000-10,000 km/s and when the expanding shell eventually reaches us then they would retain most of this energy.

Certainly, over the vastness of geological time, the Earth must have encountered supernova debris. However, the safe notion that they are a danger distant in both time and space must be dispelled. Astronomers may say, with some validity, that no comet travelling at supernova or even interstellar velocities has ever been observed. Yet neither has a supernova occurred in our galactic neighbourhood since that observed in the pre-telescopic era by Kepler in 1604; before that, they occurred in 1572, 1054, 1006, 393 and 185 giving us an average rate of one about every 333 years in our part of the galaxy. Any ejecta from these explosions will take hundreds of thousands of years to reach us, but it is not these recent supernovae that we should fear – rather, those that exploded millions, even tens of millions of years ago!

It is reasonable therefore, based on the known frequency of supernovae, to suppose that the expanding shells of very ancient supernovae should cross the solar neighbourhood at intervals of 300-500 years. Only during these brief episodes would we encounter comets and meteors of interstellar origin. If just one in ten of such events give rise to a collision

Figure 10.1 Multiple Suns in the Sky.
A shock-front of fast comets and meteors, ejected by an ancient supernova, arrives at the solar system. With velocities in excess of 10,000 km/sec even a single collision could cause the Earth to wobble on its axis. The comet-shower would cross the Earth's orbit in less than a day.

with the Earth then it has the potential to cause a major catastrophic event every three to five millennia. Whatever estimate you take, this would make impacts by supernova debris potentially more frequent than those of solar comets and asteroids.

Would an interstellar 'comet' appear any different to a solar comet? There is no physical evidence to draw on. Should we even call it a comet? One may postulate that they would be much younger than solar comets and should therefore retain short-lived radioactive isotopes and heavy elements such as Plutonium. With velocities of the order of 8,000–10,000 km/s they would streak across the orbit of the Earth in less than a day; and they would cross the entire solar system in about twelve days. Even if one of these passed close by the Sun, there might be insufficient time for it to form a bright tail. Astronomers might never see them. An asteroid travelling this fast across the sky could be mistaken for a shooting star on a photographic plate. It is certainly reassuring that astronomers should catalogue the near-earth asteroids and be on the lookout for any comets that might hit us – but if a piece of stellar debris decides to fall to earth, then we may, quite literally, never know what hit us!

The energy expended in an impact increases with the mass of the impacting body, but with the *square* of its velocity. Therefore, any given impact crater might equally be formed by a slow, massive meteor, or by a small one travelling very fast. How would we distinguish between these alternatives?

The impact of a slow, heavy asteroid or comet would deposit most of its material on the surface or in the atmosphere. Even ejecta thrown into space would ultimately fall back again. This seems to be what happened seventy million years ago at Chicxulub, which left a layer of iridium-rich debris world-wide.

By contrast, the impact of a small fast interstellar 'comet' would be more like colliding with a packet of pure energy, especially if the impact were just a glancing blow. For a total energy equivalent to a slow massive body, much less dust would be injected into the atmosphere. Indeed, if the impact were oblique, the projectile might retain enough momentum to escape not only from the Earth, but also from the solar system. This is surely the best candidate for a catastrophe in human prehistory.

At supernova velocities, a comet as small as a few kilometres in diameter has enough energy to tilt the axis of a planet or to alter the day by the amount suggested. An oblique impact from such a comet could affect the Earth's rotation, yet still be small enough and fast enough to have left no physical evidence.

Equally possible is an encounter with a series of small hyper-velocity fragments of supernova debris. The resultant effect on the angular momentum would be cumulative, equivalent to the strike of a single larger body. A near-tangential strike in the ocean might reasonably carry large volumes of ocean water and fragments of the seabed into space at sub-orbital velocities; this water would then return to the surface, in the form of a worldwide deluge of rain. Meanwhile, the comet-like body that caused all the damage, or the fragments of it, would simply disappear into interstellar space.

Tektites

There remains one possible candidate for hard evidence – *tektites*. These small heat-processed pieces of silica – melted igneous rock – may be found in various parts of the world. They occur only in 'strewn fields', of various ages, in Australia, Indonesia, West Africa; and in the USA, where they are found in Texas and Georgia. Some contain traces of nickel-iron, characteristic of meteorites. There is little else to suggest a non-terrestrial origin.

Tektites appear to be pieces of molten terrestrial rock that have been thrown high into space and then re-entered the atmosphere at high velocity; as is suggested by their streamlined disk, teardrop and dumbbell shapes. Upon examination, some tektites contain trapped bubbles of air indicative of the atmospheric composition some twenty miles up. The American geologist Dr Dean Chapman was able to reproduce the tektite forms in the laboratory, leaving little doubt that they are produced by the heat ablation of atmospheric re-entry.

Tektites are just what would be expected from a high-energy low-mass impact, which would eject terrestrial rock as it forms a crater, yet contain comparatively little material from the projectile itself. All of the known strewn-fields are geologically recent, the oldest dating from the Oligocene and Miocene, but most interesting of all are the *Australites*, found in a broad field across the whole of Australia and South East Asia. Unlike the other fields, the tektites in Australia and Indonesia cannot be associated with any known crater.

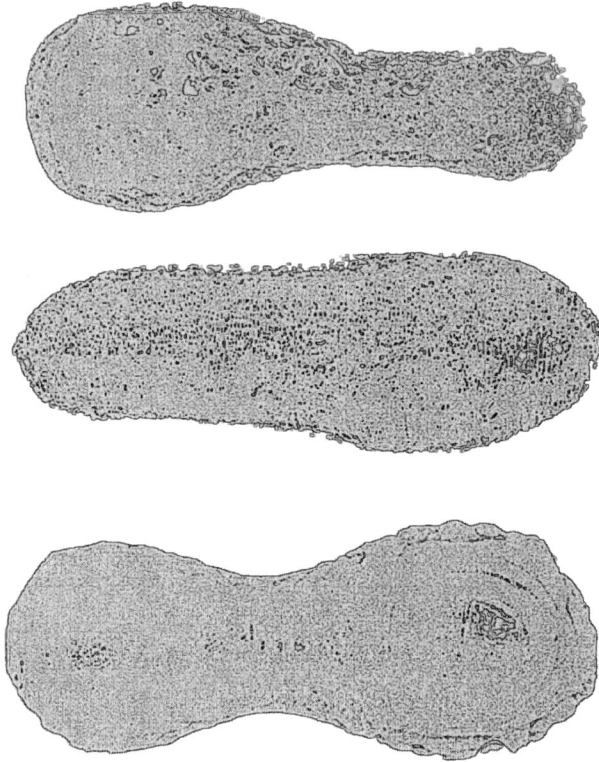

Figure 10.2 Tektites
Some typical tektite forms from Australia and Indonesia. The various teardrop shapes reveal the heat of re-entry as fragments of terrestrial rock are thrown high into space by a cometary collision or some other explosive event.

The preponderance of tektite finds across southern Australia and Tasmania would point to a strike in the Southern Ocean, between Australia and Wilkes Land, or just possibly on the Antarctic continent itself. A few Australites were discovered near Port Campbell, Victoria, embedded within deposits that were no more than 5,000 years old – an indicator that they fell at around that time.[8] Here then, may be some hard evidence of a 5,000-year-old comet impact that has left no other trace of its passing.

Dark Forces!

Finally, we must consider one further possibility: that the Earth may, from time-to-time, be influenced by cosmic forces that are, as yet, unknown to science. Recently, some pundits have declared that scientists may soon arrive at a theory to explain everything in the universe. Whenever you hear such pronouncements, it's probably a good time to take cover!

Astronomers and cosmologists recognise many violent phenomena in the universe, of which our present state of knowledge remains woefully inadequate. The gas jets that are expelled from the poles of collapsing stars are observed to reach almost to the speed of light itself and must propel away any small meteorites in their path at similar relativistic velocities. It is worth noting that: even a meteor little bigger than a football, travelling at 90% of the speed of light, would possess enough energy to tilt a planet's axis.[9] Yet, what kind of impact scar would it leave? Did perhaps, a speck of cosmic dust like this strike Siberia in 1908? [10]

Mysterious gamma-ray bursts signify huge explosions in the distant cosmos. The collapse of stars and even whole galaxies into black holes can generate gravity waves – distortions of space-time itself – that propagate at the speed of light. What happens when one of these passes through the Earth; and how often do they reach us?

For most of the twentieth century, cosmologists believed that the universe expanded from a cosmic 'egg', a singularity, and would one day collapse back under its own gravity. Some cosmologists now believe that there is just not enough mass in the visible universe to explain the observed rate of cosmic expansion. This has led to a theory that as much as nine-tenths of the universe may consist of a mysterious 'dark matter' that possesses mass, yet cannot be seen and may not even interact with ordinary atoms.

Furthermore, astronomers studying the red-shifts of distant galaxies have discovered that far from slowing-down, the expansion of the universe is actually speeding-up. As well as the dark matter there must also be a mysterious 'dark energy', a kind of anti-gravity that is propelling the galaxies apart at an increasing rate. If we have dark matter and dark energy then, presumably, there must also be 'dark forces' at work in the universe![11]

One is bound to ask: where is all this unseen matter and energy? Why do we not find any of this mysterious stuff in the vicinity of our own safe little world? And what happens when we do encounter it?

An Encounter with Shadow Earth!

If you thought the physics of the Earth's rotation was complex then wait until you sample some of the latest ideas in cosmic physics! One current theory as to where the missing mass of the universe lies is that it may not be in our universe at all, but in a parallel, or 'shadow' universe.

At the heart of modern cosmology lies the attempt to unify the force of gravity with the other fundamental forces known from particle physics: electromagnetism and the strong and weak nuclear forces, in a so-called quantum theory of gravity. The quantum world of the atom has some very strange properties that defy everyday common sense, such as the uncertainty principle; or pairs of particles that may pop in and out of existence from the vacuum of space. One day soon, some new 'Einstein' may devise a theory of quantum-gravity, bringing together Relativity theory that governs the very large, and the quantum physics that rule the sub-atomic domain. This raises the intriguing possibility that some strange and totally unforeseen quantum effects may at times occur in the macro world.

Current thinking, if one may cite Professor Stephen Hawking in his book *The Universe in a Nutshell,* is that we inhabit, not a single universe, but an infinity of finite universes: a multiverse; each universe may have completely different laws of physics in which anything may be possible.[12] Moreover, the four dimensions of space-time that we inhabit may not be all there is; the latest theory calls for no less than *eleven* dimensions. Most of these extra dimensions are rolled-up within atomic particles so that we cannot see them.

However, one intriguing possibility is that at least one of these higher dimensions is flat like those we live in; and that we inhabit only a four dimensional layer of this higher-dimensional space, which the physicists call a *brane*-world. Hawking considers that these branes may be separated from each other by no more than millimetres (as measured in a fifth dimension) and that a mass lying in an adjacent brane may exert its gravity in our own universe.

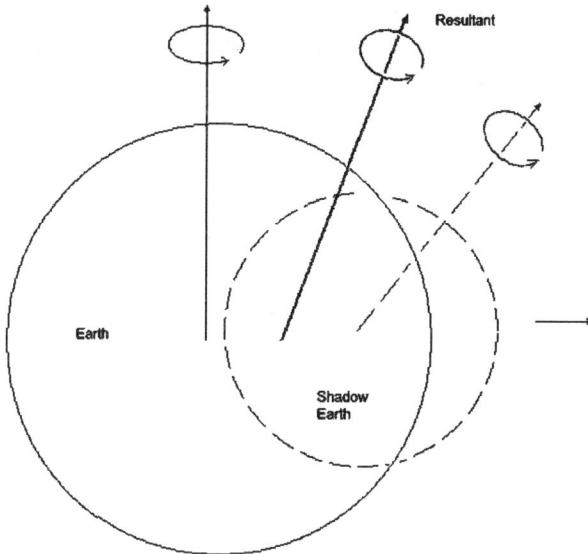

Figure 10.3 An Encounter with Shadow Earth
A massive planet in a parallel universe temporarily occupies the same space-time co-ordinates as
our Earth, but is separated from us by the short distance that separates the universes in an
unseen fifth dimension. To us it would appear that the earth's centre of gravity had shifted and
(since the axis of rotation must always pass through the centre of gravity) it would trigger a
wobble of the Earth's axis. If the shadow mass were rotating at a different rate, or even retrograde,
then this also has the capacity to alter the length of day.

The force of gravity extends to infinity in our three dimensions, so why should it not do so in all eleven? However, the influence of such gravity anomalies could only extend a few centimetres otherwise, as Hawking points out; we would have quite different laws of physics: planets would spiral into their suns and atoms would be unstable. It seems we should only feel the influence of a shadow mass from an adjacent brane-world when the anomaly passes right through us.

Now consider this for a moment. Most of our universe is empty space and so the same would probably be true for our nearest parallel universe. All we would notice in our own universe would be a gravity 'anomaly' – a point source of gravity emanating from apparently empty space. Gravity anomalies are the stuff of science fiction writers - but they don't have to explain what could cause them! Probably we would only know for certain that they exist when our own spaceships venture far out and find themselves pulled off course.

How might we experience such a strange phenomenon? Consider that a shadow-earth orbits a shadow-sun and for a short time it occupies the same four dimensions of space-time as ourselves, yet separated from us only by a very short distance in the fifth dimension. It would appear as if the Earth's centre of gravity had momentarily stepped-out and then moved back again. For an instant, the same gravity that pulls us down would pull us sideways or even upwards – a bit like being on a fairground ride. The effect would last just a few minutes or even just seconds – only as long is takes for an anomaly travelling at say, 30km/sec to traverse the Earth's diameter.

Most interesting from the point of view of the present author's hypothesis is that a gravity anomaly has the potential to cause Noah's Flood. It could excite the core wobble without leaving any scar on the Earth's surface. All the effects that have previously been described would then unfold. Also, alone among the theories so far discussed, it could perhaps make the Earth's rotation stand still or pause temporarily – raising the possibility that the Biblical references to such an event could indeed have a physical cause after all.

Here surely, are cosmic cannonballs enough to cause changes in the Earth's rotation. Those who would continue to assert that such things cannot happen have learned nothing from the history of science.

If a catastrophic event, of whatever nature, really has occurred during human prehistory then the statistical probabilities, cease to be of any further relevance. Only two such events: one at the close of the ice age and another during the mid-Holocene, would suffice to explain most of the phenomena that we find in the physical and mythological record. Lesser irregularities could then be explained by intermediate-scale impact events, or by exceptional volcanic eruptions.

Sometimes, to paraphrase *Sherlock Holmes*: when you have ruled out the probable and the impossible, the improbable explanation is the only one that remains!

References to Chapter Ten

1 Alvarez (1981)

2 Gersonde et al (1997).

3 I have examined this in more detail in chapter 10 of *The Atlantis Researches*.

4 This is not unlike the theories of Velikovsky! However in this case the comet would not actually originate from Jupiter, but might appear so to an Earth-based observer.

5 See for example Weissman (1982).

6 This is actually a type II supernova. Type I occurs in the total destruction of one partner of a double star.

7 Charles, P.A & Sewa, F.D. 1995, pp 62-63. For example, the so called Cygnus 'superbubble' is believed to have been formed by the evolution and explosion of as many as 100 separate supernovae over the course of the past 3 million years.

8 Gill, E.D. (1970).

9 If you enjoy science-fiction then consider this possibility. An alien spacecraft is set on course for Earth from a nearby star system. After accelerating to half the speed of light, the crew are killed by some mishap and the craft fails to decelerate. It therefore cruises on autopilot until it slams into the Earth. At half the speed of light it would possess quite enough energy to cause Noah's flood.

10 On the absence of debris from the Tunguska impact, see Svertsov, 1996.

11 Consider, for example, a strongly magnetic force that could exert a torque on the iron core, triggering a wobble, yet without leaving a scar on the surface. However, there is currently no known astronomical phenomenon that could do this.

12 Hawking, S. 2001, p178-188.

11

A World Reborn!

We come full circle, from mythology to science and back to mythology. The traditions of a Great Flood in human prehistory could after all be a real memory of an ancient cosmic encounter.

However, in summarising and drawing conclusions from the preceding chapters, an author must appreciate that this brief final summary is the only chapter that some people, especially book reviewers, will actually read. Many others, even those who have understood the arguments and the pattern of evidence, still will not accept the conclusion – that some of the most basic parameters of our everyday world may have changed only a few thousand years ago.

Yet there is nothing in the preceding chapters that violates either physical laws or the current state of knowledge in any of the disciplines addressed. Nothing is speculative and all the references cited are sound. There is a saying in English that says to someone with ideas beyond their capacity: "keep your feet on the ground" – but have you ever been in an earthquake? What if the very ground itself is not safe to stand on? What if the sky really could fall on our heads, just as the ancient Celts feared? What if the sea really could rise-up unexpectedly? What if the length of the day and the seasons really *can* change within a human lifetime?

We may deduce that the Earth has probably been struck by comets and asteroids, and possibly by more exotic assailants, at regular intervals throughout geological time. At intervals of several thousand years, we are struck by a fragment of low-mass supernova ejecta, travelling at a very high velocity, which reconfigures the rotation and climate of the planet. Cumulatively, these impacts may even alter the orbit. Such an event was the Great Flood, of which our mythologies and religions tell us.

Although these impacts may be globally devastating, they do not destroy all life on Earth, nor even result in mass extinctions. The impacting bodies are so small and energetic that they may bounce off the Earth and most of the ejecta will be carried into space. Devastation and loss of species will be only localised. When the poles shift, there are always two neutral zones around 90° from the meridian of greatest effect, where life can survive. Although some species may be lost, after a few centuries even the devastated parts of the planet would be completely repopulated. The outcome would be conceivably little worse than the recovery from a major volcanic episode.

However, at much longer intervals, separated by millions of years, the Earth is struck by a solar asteroid or comet. These are of higher mass and low velocity and they deposit most of this mass onto the surface and in the atmosphere. They do not possess enough energy to affect the rotation of the planet, but the resulting dust veil is world wide and it causes mass extinctions of species. When the dust settles, life has to re-evolve again from a few chance survivals. Such an event was the Chicxulub impact and probably we have not suffered an impact of this type in the past few million years. It may be that the so-called mass extinction events have occurred whenever the two types of impact have occurred together, within a relatively short time.

However, during the geologically recent period that we call the Ice Age, there have probably been many of the high-energy events, which have punctuated the long-term cycles of climate change determined by the orbital perturbations.

As we approach the period of human civilisation, we may detect the occurrence of two, possibly three or more unexplained events that may have given rise to our Flood and Deluge myths. The earliest of these occurred at the Younger Dryas event, around 8000 BC. Another may have occurred around 5500 BC, with the most recent around 3100 BC at the dawn of history. While the most recent encounter gave rise to the strongest memory, it seems likely that our myths remember a composite of them all, with some religions remembering the separate ages of mankind that were punctuated by these events. Less globally devastating impacts are also a possibility, around 4400 BC, 1650 BC and AD 540, but these might equally be recording volcanic dust veils.

At what point in prehistory human society first devised a calendar must be forever unknown. From simple origins, calendar knowledge evolved with the needs of farming societies. The very conservatism of calendar tradition may thus have preserved for us some of the characteristics of the world as it was in an earlier age.

The myths suggest that before the Flood there was a golden age, when the climate was warm and equable. This corresponds with what we know of the climate during mid-Neolithic times up to 5,000 years ago. The pattern of climate and calendrical evidence suggests that the obliquity may then have been less, perhaps only 20-22° and the length of day longer, at 360 days per year. Many of the world's oldest civilisations, in China, India, Central America, Babylon, Egypt (and one should include megalithic Europe) all grew up at this era. These societies developed sophisticated astrology and religious cosmologies; and they all devised calendars based on a 360-day year. There may also have been a large and prominent short-period comet in the skies during this period, which slowly broke up, giving rise to beliefs in sky gods and demi-gods.

Then, at some time around 3100 BC, the sky suddenly filled with bright, fast moving comets and meteors as the expanding shell of an ancient supernova passed through the solar system. It must have seemed to our ancestors that the gods were fighting among themselves and the various stories of sky gods battling with giants, dragons and serpents in the sky may date from this time.

One of these fast interstellar comets struck the Earth, probably a glancing blow near one of the poles, or in the ocean. It permanently altered the length of the day and caused a tilt of the axis. Another legacy was a geographical pole-shift of a fraction of a degree, just enough to cause permanent sea level changes around the world. The comet itself bounced off and left the solar system altogether.

The seawater that the impact had thrown up returned to the surface as a world-wide deluge. The impact excited the Chandler wobble and briefly, pole tides submerged many low-lying coastal areas. Thus, the event was remembered by the scattered survivors as both a *deluge* from heaven and as a *flood* from the sea. After twenty years, the Chandler wobble was damped and the Great Flood was over, but probably its most damaging effects had passed within less than a year, as the solid Earth adjusted its figure. It left behind a permanent legacy of sea level and climatic changes.

The impact also triggered the Earth's core wobble (the nearly-diurnal wobble). Although less globally devastating, its effects persisted for much longer, perhaps until 500 BC. Throughout the third millennium BC it would have been noticeable to the naked eye as a spiralling motion of the whole sky with a period of 432 days. These combined with the annual seasons to cause seven-year rhythms in the climate. In India and Babylon, and probably elsewhere too, this was built into the calendar and entered into the religious material. By about 1500 BC, the motion would have become too small to detect with the naked eye and it was forgotten.

Sometime during this period, probably around 2900-2700 BC our ancestors recognised that their old 360-day calendars no longer worked. They began to adopt 365-day solar calendars, or else, they resorted to lunar calendars. Certainly, it led to an expansion of astrological measurements, as people all over the world tried to predict the portents of their gods in the sky. Just a few generations after the Flood, the threat must have seemed very real. This may explain why people in Egypt, Neolithic Europe and elsewhere were motivated to build astronomically aligned structures such as the pyramids and the stone circles; and why ancient people could achieve a standard of astronomy that was not exceeded until the invention of the telescope.

We may see though, that earlier megaflood events must have been even more devastating. Certainly that at the close of the Ice Age, which melted the North American ice sheets and contributed to the extinction of the mammoths and other ice-age macro fauna. We should not even rule out a somewhat larger excursion of the poles at that time, just as the nineteenth century commentators had proposed, although its cause may be not quite what they thought.

Yet, the legacy has not been entirely a negative one. These events shaped our modern world and without the spur to evolution that extinctions brought about, humans would not be here at all. Every chance event creates both winners and losers and we alive today are all the winners in the game of cosmic chance. Not only did our own ancestors survive the Great Flood, they have survived *all* of the thousands of floods that preceded it. One of your mammalian ancestors even survived the great extinction of the dinosaurs. And you thought you were unlucky!

Modern astronomers and scientists may look up at the sky and understand the causes of the threat. Our ancestors could only see it all as the will of omnipotent sky gods. They may have thought that religious rites and sacrifices could appease their malevolent deities. Modern astronomers may have begun to chart the comets and near-earth asteroids and predict their returns; but if a comet approaches us from interstellar space on a hyperbolic trajectory then they probably wouldn't have time even to calculate the orbit before it hit us.

Within the next few generations we may perhaps foresee the development of space technology and astronomy to a point where humanity can protect itself from the comet and asteroid threat. There have been various suggestions as to how we might deflect or destroy an asteroid before it hits us – but such expeditions would require many months or even years of planning. We must rely on the vigilance of the astronomers.

To deflect a comet of cosmic origin however, implies a capability for interstellar space flight and the availability of truly colossal sources of energy. Furthermore the threat would have to be detected long before it reaches our solar system. We may be centuries away from the kind of routine interstellar travel that science fiction writers conceive, but we may hope that even this will one day be within human capability. There can surely be no greater spur to the development of space technology than to prevent the next Great Flood.

Or perhaps the ancients had the right idea after all and prayer may be the better course. It may never happen. So let us eat, drink and be merry!

Appendix A

A Reconstruction of the Coligny Calendar

This reconstruction of the Celtic Calendar of Coligny is based upon the principle of alternating five and six year periods. Each five-year cycle is constrained by the rules of the extant Coligny fragment. The variable month Equos is reconstructed according to the system of MacNeil, with alternately 28 or 30 days, giving a total of 1831 days for the five years.[1] This follows the premise that it was classified a 'not good' month. The hypothetical six-year period has the variable month reconstructed with 29 days, but with the first month increased to 30 days. Clearly, there could be many other ways to configure the six-year periods, without violating the rules of the extant five-year period.

The reconstruction shown here is strictly lunar. Each eleven-year cycle totals 4016 days and the thirty-year period is therefore alternately 10955 or 10956 days, totalling 21911 days for the sixty years. If instead it is desired to hold the calendar to solar time then four extra days need to be interpolated, to bring the total up to the 21915 days approximating sixty solar years.

References

1   MacNeill E, (Dublin 1928), On the Notation and Caligraphy of the Calendar of Coligny, Eriu, X, (1926-8), VI.

A Reconstruction of the 'Celtic' Calender

Year No.	Inter-calary month	Month 1 SAMON (JUN)	Month 2 DVMANN (JUL)	Month 3 RIVROS (AUG)	Month 4 ANAG'NT (SEP)	Month 5 OGRON (OCT)	Month 6 CVTIOS (NOV)	Inter-calary month	Month 7 GIAMON (DEC)	Month 8 SIMIVIS (JAN)	Month 9 EQVOS (FEB)	Month 10 ELEMBIV (MAR)	Month 11 EDRIN (APR)	Month 12 CANTLOS (MAY)	Days/ year	Days/ period
1	30	30	29	30	29	30	30		29	30	30	29	30	29	385	
2		30	29	30	29	30	30		29	30	28	29	30	29	353	
3		30	29	30	29	30	30	30	29	30	30	29	30	29	385	
4		30	29	30	29	30	30		29	30	28	29	30	29	353	
5		30	29	30	29	30	30		29	30	30	29	30	29	355	
6	30	30	29	30	29	30	30		29	30	30	29	30	29	385	
7		30	29	30	29	30	30		29	30	29	29	30	29	354	
8		30	29	30	29	30	30		29	30	29	29	30	29	354	
9	30	30	29	30	29	30	30		29	30	29	29	30	29	384	
10		30	29	30	29	30	30		29	30	29	29	30	29	354	
11		30	29	30	29	30	30		29	30	29	29	30	29	354	4016
12	30	30	29	30	29	30	30		29	30	30	29	30	29	385	
13		30	29	30	29	30	30		29	30	28	29	30	29	353	
14		30	29	30	29	30	30	30	29	30	30	29	30	29	385	
15		30	29	30	29	30	30		29	30	28	29	30	29	353	

(Continued)

A Reconstruction of the 'Celtic' Calender (*Continued*)

Year No.	Inter-calary month	Month 1 SAMON (JUN)	Month 2 DVMANN (JUL)	Month 3 RIVROS (AUG)	Month 4 ANAG'NT (SEP)	Month 5 OGRON (OCT)	Month 6 CVTIOS (NOV)	Inter-calary month	Month 7 GIAMON (DEC)	Month 8 SIMIVIS (JAN)	Month 9 EQVOS (FEB)	Month 10 ELEMBIV (MAR)	Month 11 EDRIN (APR)	Month 12 CANTLOS (MAY)	Days/ year	Days/ period
16		30	29	30	29	30	30		29	30	30	29	30	29	355	
17	30	30	29	30	29	30	30		29	30	30	29	30	29	385	
18		30	29	30	29	30	30		29	30	29	29	30	29	354	
19		30	29	30	29	30	30		29	30	29	29	30	29	354	
20	30	30	29	30	29	30	30		29	30	29	29	30	29	384	
21		30	29	30	29	30	30		29	30	29	29	30	29	354	
22		30	29	30	29	30	30		29	30	29	29	30	29	354	4016
23	30	30	29	30	29	30	30		29	30	30	29	30	29	385	
24		30	29	30	29	30	30		29	30	28	29	30	29	353	
25		30	29	30	29	30	30	30	29	30	30	29	30	29	385	
26		30	29	30	29	30	30		29	30	28	29	30	29	353	
27		30	29	30	29	30	30		29	30	30	29	30	29	355	
28	30	30	29	30	29	30	30		29	30	30	29	30	29	385	
29		30	29	30	29	30	30		29	30	29	29	30	29	354	
30		30	29	30	29	30	30		29	30	29	29	30	29	354	2924

Totals of first 30 yr cycle = **10956**

[Note: 30 years falls halfway through a six-year cycle]

(Continued)

261

Second Cycle *(Continued)*

Year No.	Inter-calary month	Month 1 SAMON (JUN)	Month 2 DVMANN (JUL)	Month 3 RIVROS (AUG)	Month 4 ANAG'NT (SEP)	Month 5 OGRON (OCT)	Month 6 CVTIOS (NOV)	Inter-calary month	Month 7 GIAMON (DEC)	Month 8 SIMIVIS (JAN)	Month 9 EQVOS (FEB)	Month 10 ELEMBIV (MAR)	Month 11 EDRIN (APR)	Month 12 CANTLOS (MAY)	Days/year	Days/period
31	30	30	29	30	29	30	30		29	30	29	29	30	29	384	
32		30	29	30	29	30	30		29	30	29	29	30	29	354	
33		30	29	30	29	30	30		29	30	29	29	30	29	354	1092
34	30	30	29	30	29	30	30		29	30	30	29	30	29	385	
35		30	29	30	29	30	30		29	30	28	29	30	29	354	
36		30	29	30	29	30	30	30	29	30	30	29	30	29	385	
37		30	29	30	29	30	30		29	30	28	29	30	29	353	
38		30	29	30	29	30	30		29	30	30	29	30	29	355	
39	30	30	29	30	29	30	30		29	30	30	29	30	29	385	
40		30	29	30	29	30	30		29	30	29	29	30	29	354	
41		30	29	30	29	30	30		29	30	29	29	30	29	354	
42	30	30	29	30	29	30	30		29	30	29	29	30	29	384	
43		30	29	30	29	30	30		29	30	29	29	30	29	354	
44		30	29	30	29	30	30		29	30	29	29	30	29	354	4016
45	30	30	29	30	29	30	30		29	30	30	29	30	29	385	

(Continued)

Second Cycle *(Continued)*

Year No.	Inter-calary month	Month 1 SAMON (JUN)	Month 2 DVMANN (JUL)	Month 3 RIVROS (AUG)	Month 4 ANAG'NT (SEP)	Month 5 OGRON (OCT)	Month 6 CVTIOS (NOV)	Inter-calary month	Month 7 GIAMON (DEC)	Month 8 SIMIVIS (JAN)	Month 9 EQVOS (FEB)	Month 10 ELEMBIV (MAR)	Month 11 EDRIN (APR)	Month 12 CANTLOS (MAY)	Days/ year	Days/ period
46		30	29	30	29	30	30		29	30	28	29	30	29	353	
47		30	29	30	29	30	30	30	29	30	30	29	30	29	385	
48		30	29	30	29	30	30		29	30	28	29	30	29	353	
49		30	29	30	29	30	30		29	30	30	29	30	29	355	
50	30	30	29	30	29	30	30		29	30	30	29	30	29	385	
51		30	29	30	29	30	30		29	30	29	29	30	29	354	
52		30	29	30	29	30	30		29	30	29	29	30	29	354	
53	30	30	29	30	29	30	30		29	30	29	29	30	29	384	
54		30	29	30	29	30	30		29	30	29	29	30	29	354	
55		30	29	30	29	30	30		29	30	29	29	30	29	354	4016
56	30	30	29	30	29	30	30		29	30	30	29	30	29	385	
57		30	29	30	29	30	30		29	30	28	29	30	29	353	
58		30	29	30	29	30	30	30	29	30	30	29	_30_	29	385	
59		30	29	30	29	30	30		29	30	28	29	30	29	353	
60		30	29	30	29	30	30		29	30	30	29	30	29	355	1831

56 synods of Saturn = 56 x 378.0928 = 21173.1968 days

717 months = 717 x 29.53059 = 21173.43303 days

Totals of second 30 yr cycle = **10955**

Totals for 60 years = **21911**

*717 calendar months = 21174 days

11 year cycle: 1832+2184 = 4016 days

Lunar solar difference 30 years less 371 months = 1.41785 days

Accuracy allowing 1 day solar drift each 30 years = 30/0.41785 = 1 day in 71.80 years

Appendix B

The Coligny Calendar – an Alternative Reconstruction

This second reconstruction is similarly based upon the principle of alternated five and six year periods. However, in this case an extra day has been added to the variable month in the first eleven-year cycle and two days to the second eleven-year cycle. One consequence of this configuration is that the start of a month falls behind the real Moon by 3 days every 22 years. After 44 years it would therefore give the situation described by Pliny, who tells us that the month began on the sixth day after new moon.

This would hold the lunar cycle naturally in balance with the solar cycle over a period of twenty two years, with a discrepancy of only about twelve minutes. This compares most favourably with the two-hour discrepancy of the Meton cycle over nineteen years. A further advantage is that a simple cycle of 5 + 6 years repeats indefinitely, with no other complex rules to remember.

However, we know from the sources that there was also a thirty-year cycle based upon direct observation of Saturn. If there were no other intervention, then the thirty-year and eleven-year cycles would repeat every 330 years.

A Reconstruction of the 'Celtic' Calender based on a pure 11-year cycle

Year No.	Inter-calary month	Month 1 SAMON (JUN)	Month 2 DVMANN (JUL)	Month 3 RIVROS (AUG)	Month 4 ANAG'NT (SEP)	Month 5 OGRON (OCT)	Month 6 CVTIOS (NOV)	Inter-calary month	Month 7 GIAMON (DEC)	Month 8 SIMIVIS (JAN)	Month 9 EQVOS (FEB)	Month 10 ELEMBIV (MAR)	Month 11 EDRIN (APR)	Month 12 CANTLOS (MAY)	Days/ year	Days/ period	Months/ period
1	30	30	29	30	29	30	30		29	30	30	29	30	29	385		
2		30	29	30	29	30	30		29	30	29	29	30	29	354		
3		30	29	30	29	30	30	30	29	30	30	29	30	29	385		
4		30	29	30	29	30	30		29	30	29	29	30	29	354		
5		30	29	30	29	30	30		29	30	30	29	30	29	355	1833	62.071
6	30	30	29	30	29	30	30		29	30	29	29	30	29	384		
7		30	29	30	29	30	30		29	30	29	29	30	29	354		
8		30	29	30	29	30	30		29	30	29	29	30	29	354		
9	30	30	29	30	29	30	30		29	30	29	29	30	29	384		
10		30	29	30	29	30	30		29	30	29	29	30	29	354		
11		30	29	30	29	30	30		29	30	29	29	30	29	354	2184	73.957
12	30	30	29	30	29	30	30		29	30	30	29	30	29	385		
13		30	29	30	29	30	30		29	30	29	29	30	29	354		
14		30	29	30	29	30	30	30	29	30	30	29	30	29	385		
15		30	29	30	29	30	30		29	30	30	29	30	29	355		
16		30	29	30	29	30	30		29	30	30	29	30	29	355	1834	62.105
17	30	30	29	30	29	30	30		29	30	29	29	30	29	384		
18		30	29	30	29	30	30		29	30	29	29	30	29	354		

(Continued)

A Reconstruction of the 'Celtic' Calender based on a pure 11-year cycle (*Continued*)

Year No.	Inter-calary month	Month 1 SAMON (JUN)	Month 2 DVMANN (JUL)	Month 3 RIVROS (AUG)	Month 4 ANAG'NT (SEP)	Month 5 OGRON (OCT)	Month 6 CVTIOS (NOV)	Inter-calary month	Month 7 GIAMON (DEC)	Month 8 SIMIVIS (JAN)	Month 9 EQVOS (FEB)	Month 10 ELEMBIV (MAR)	Month 11 EDRIN (APR)	Month 12 CANTLOS (MAY)	Days/ year	Days/ period	Months/ period
19		30	29	30	29	30	30		29	30	29	29	30	29	354		
20	30	30	29	30	29	30	30		29	30	29	29	30	29	384		
21		30	29	30	29	30	30		29	30	29	29	30	29	354		
22		30	29	30	29	30	30		29	30	29	29	30	29	354	2184	73.957

First 11 year cycle: 1833+2184 = **4017** days

Second 11 year cycle 1834+2184 = **4018** days

Lunar solar difference: 22 years less 272 months = **3.00819** days

Totals of 22 year cycle = **8035   272.091**

266

Appendix C

Climate Zones of the European Holocene

The first attempts to classify the phases of climate since the end of the Ice Age were led by the Scandinavian pioneers Blytt and Sernander, based on the phases of faster and slower peat growth in north European peat bogs. Iversen would later modify this with an early Holocene warm-phase, (up to about 5000BP) followed by a late Holocene cool phase. Later researchers would then classify a series of pollen zones identifying the principal vegetation cover within these broader classifications.

More modern research failed to extend these climate zones beyond Europe and the Blytt-Sernander zones therefore tend to be used only in a colloquial sense. However, it is clear that the Climatic Optimum up to 5000 BP was real throughout the northern hemisphere and the sub-phases still represent a broad summary of the climate changes for Europe.

Climate Zones of the European Holocene

Climate Zone	Approx. age in years BP	Climate Type	Peat Type	Principal Vegetation	Archaeology
Pre–Boreal	10 000–9500	cool/ dry		tundra	Paleolithic
Boreal	9500–7000	warm/ dry	humified peat	birch/ pine forests	Paleolithic/ Mesolithic
Atlantic	7000–5000	warm/ wet	unhumified peat	elm/ lime forests	Mesolithic/ early and mid– Neolithic
Sub–Boreal	5000–2500	warm/ dry	humified peat	oak/ grasses/ agriculture	Late Neolithic and Bronze Age
Sub–Atlantic	2500–present	cool/ wet	unhumified peat	agriculture	Iron Age

Bibliography

A

Airy, G. B. *Change of Climate*, Atheaeum, 1717, 384-5 (1860)

Aitken, M. J. *Physics and Archaeology*, Clarendon Press, Oxford (1974)

Agassiz, L. (1840) *Etudes sur les Glaciers* (in English translation and introduction by A.V. Carozzi), Hafner, New York (1967)

Aldred, C., *Akhenaten, King of Egypt*, London (1988)

Allen, C.W. *Astrophysical Quantities*, The Athlone Press, London, (1977)

Allen, D. S. & Delair, J. B., *When the Earth Nearly Died*, Gateway Books, Bath (1995)

Allen, J. M. *Atlantis: The Andes Solution*, Windrush Press, Moreton-in-Marsh (1998)

Alvarez, L.W. et al, *Extraterrestrial cause for the Cretaceous-Tertiary Extinction: Science*, 208, 1095-1105 (1980)

Alvarez, L.W., *Current status of the impact theory for the terminal Cretaceous extinction*, in *Geological implications of impacts of large asteroids and comets on the earth* (eds L.T. Silver & P.H. Schulz), Geol. Soc. Am. Spec. Pap., 190, 305-315 (1982)

Aveni, A, *Stairways to the Stars – Skywatching in Three Great Ancient Cultures*, John Wiley & Sons, London (1997)

B

Babbitt, F. C. *Plutarch's Moralia*, Vol V, William Heineman Ltd, London. (1936)

Baillie, M. G. L. Irish oaks record prehistoric dust veils drama, Archaeology Ireland, 2(2): 71-4. (1988)

Baillie, M. G. L. and Munro, M. A. R. *Irish tree rings, Santorini and volcanic dust veils, Nature*, 332, March (1988)

Baillie, M. G. L. *Do Irish Bog Oaks Date the Shang Dynasty?*, Current Archaeology, 117, 310-313. (1990)

Baillie, M. G. L, *Extreme environmental events and the linking of the tree-ring and ice-core records*, J.S. Dean et al (ed) *Tree Rings, Environment and Humanity* (1994)

Bailey, M. E., Clube, S.V.M. & Napier, W.M., *The Origin of Comets*, Pergamon Press, Oxford (1990)

Barber, D. C. et al, *Forcing of the cold event of 8,200 years ago by catastrophic drainage of Laurentide lakes*, Nature, 400, 344-8 (1999)

Bauval, R. & Gilbert, A. *The Orion Mystery*, William Heineman Ltd, (1994)

Baxter, J. & Atkins, T. *The Fire Came By*, Macdonald & Jones Publishers Ltd, London. (1976)

Bayliss, A. et al, *Precise dating of the Norfolk timber circle*, Nature, 492 (1999)

Beaton, J. M. *The transition on the coastal fringe of Greater Australia*, Antiquity, 69, 798-806 (1995)

Bell, B. *The Oldest Records of the Nile Floods*, Geog. J., 136, pp569-573 (1970)

Bell, B. *The Dark Ages in Ancient History: The First Dark Age in Egypt*, Am. J. Archaeol, 75, 1-26 (1971)

Bickerman, E. J. *Chronology of the Ancient World*, Thames and Hudson, London (1968)

Black, R. F. & Barksdale, W.F. *The Oriented lakes of Northern Alaska*, in Journal of Geology, 57: 105-118. (1949)

Bowra, C. M. (tr), *Pindar: The Odes*, Penguin, Harmondsworth (1969)

Brennan, M., *The Stars and the Stones*, Thames & Hudson, London. (1983)

Bromwich, R. *Troiedd Ynys Prydein – The Welsh Triads*, University of Wales Press, Cardiff. (1978)

Brothwell, D. and Kzanowski, W. *Evidence of Biological differences between early British populations from Neolithic to Medieval times as revealed by eleven commonly Available cranial vault measurements*, J. Archaeol Sci, 1, 249-260. (1974)

Buckland, P. C., Dugmore, A. J. & Edwards, K. J., *Bronze Age Myths? Volcanic activity and human response in the Mediterranean and North Atlantic regions*, Antiquity 71, 581-593 (1997)

Buckland, W., *Reliquiae Diluvianae; or Observations on the Organic Remains Contained in Caves, Fissures and Diluvial Gravel and on Other Geological Phenomena Attending the Action of a Universal Deluge*, John Murray, London (1823)

Buffet, B.A., *A mechanism for decade fluctuations in the length of day*, Geophys. Res. Lett. 23, 3808-3806 (1996)

Buffet, B.A., *Geodynamic estimates of the viscosity of the Earth's inner core*, Nature, 388, 571-3 (1997)

Bullard, E. C., Ann. Rev. Earth Planet Sci. 3,1.

Burgess, C. *Volcanoes, Catastrophe and the Global crisis of the late Second Millennium BC*, Current Archaeology, 117, 325-329. (1990)

Burl, H.A.W. *The Stone Circles of the British Isles*, Yale University Press Ltd, London. (1976)

Burl, A. *From Carnac to Callanish, The prehistoric stone rows and avenues of Britain, Ireland and Brittany*, Yale, Newhaven & London (1993)

Burl, H.A.W. *Megalithic Brittany*, Thames & Hudson, London. (1985)

Burnet, T. (1690-1691) *The Sacred Theory of the Earth*. reproduced with introduction by B. Willey, Southern Illinois Univ. Press (1965)

Burstein, S. *The Babyloniaca of Berossus*

C

Campbell, J. & Eliak, M., *The Universal Myths – heroes, Gods, tricksters & others*, Penguin, Harmondsworth (1990)

Capitain, N. & Xiao, N., *Some terms of nutation derived from the BIH data*, Geophys, J. R. astr. Soc 68, 805-814 (1982)

Castleden, R. *The Stonehenge People*, Routledge and Kegan Paul, London. (1987)

Chapman, C., *Hazard to Civilization of Asteroid and Cometary Impacts*, Asteroid Hazard Conference USSR Academy of Sciences, St Petersburg, October 10 (1991)

Chapman, C.R. & Morrison, D., *Impacts on the Earth by Asteroids and Comets: assessing the hazard*, Nature, 367, 33-39 (1994)

Charles, A., & Sewa, F.D., *Exploring the X-ray Universe*, Cambridge Univ. Press (1995)

Charles, C., *The ends of an era*, Nature, 394, 422-3 (1998)

Charles, R. H. *The Book of Enoch*, SPCK (19th edition) London (1987)

Cherniss, H. & Helmbold, W. *Plutarch's Moralia, Vol XII*, William Heineman Ltd, London. (1957)

Childe, V. G. *The Dawn of European Civilisation*, 6th edition, Routledge & Kegan Paul, London. (1957)

Chu Ko-Chen *A preliminary study on the climatic fluctuations during the last 5,000 years in China*, Scientia Sinica, XVI, No.2 (1973)

Clarke, B. (ed), *The Life of Merlin – Vita Merlini*, Univ. Of Wales Press. (1973)

Clausen, H. B., Friedrich, W. L. & Tauber, H. *Dating of the Santorini Eruption*, Nature, 332, 401. (1988)

Clube, V. & Napier, W., *The Cosmic Serpent*, Faber & Faber, London, (1982)

Crawford, O.G.S. *Lyonesse*, Antiquity, 1, 5-14 (1927)

Crossley-Holland, K., *The Norse Myths*, Penguin, Harmondsworth (1980)

Cummins, W. A., *King Arthur's Place in Prehistory: The Great Age of Stonehenge*, Sutton Publishing, Stroud (1992)

D

Dahlen, F.A. *Latest spin on the core*, Nature 402, 29-9 (1999)

Dansgaard, J.W., White, W. C., & Johnsen, S.J., *The abrupt termination of the Younger Dryas climate event*, Nature, 339, 532-3 (1989)

Darwin, C. *On the Origin of Species by Natural Selection*, John Murray, London. (1859)

Darwin, Sir G. H, *On the influence of geological changes on the Earth's axis of rotation*, Phil. Trans. R. Soc, part 1, 167, 271-312 (1877)

Davies, E. *Celtic Researches*, J. Booth, London. (1804)

Davies, E. *The Mythology and Rites of the British Druids*, J. Booth, London. (1809)

de Jong, T, & van Soldt, T. *The earliest known solar eclipse record redated*, Nature, 388, 238-240 (1989)

De Selincourt, A. *Herodotus: The Histories*, Penguin Books, Harmondsworth (1954)

De Selincourt, A. *Arrian: The Campaigns of Alexander*, Penguin, Harmondsworth. (1958)

Desroches-Noblecourt, C., *Tutankhamen*, Penguin, harmondsworth (1965)

Donnelly, I. *Atlantis: The Antediluvian World*, Harper & Brothers, New York. (1882)

Donnelly, I. *Ragnarok: The Age of Fire & Gravel* (originally published 1883), University Books, New York. (1970)

Dunbavin, P. *The Atlantis Researches – the Earth's Rotation in Mythology and Prehistory*, Third Millennium Publishing, Nottingham. (1995)

Dunbavin, P. *Picts and Ancient Britons*, Third Millennium Publishing, Nottingham. (1998)

Dutt, R. C. *The Ramayana & The Mahabharata*, Everyman's Library, Dent, London.

Duval, P.-M & Pinault, G *Recueil des Inscriptions Gauloises, 3: Les calendriers (Coligny, Villards d'Heria)*. Paris: Supplement a Gallia 45 (1986)

E

Emery, W. B. *Archaic Egypt*, Penguin (1961)

Ellis, B. G., *An oblique view of climate*, Nature, 396, 405-6 (1998)

Ellis Davidson, H. R. *Gods and Myths of Northern Europe*, Penguin Books, Harmondsworth. (1964)

Ellis Davidson, H. R. *Scandinavian Mythology*, Hamlyn, London. (1982)

Eogan, G. *Knowth and the Passage-Tombs of Ireland*, Thames and Hudson, London, 1986

F

Fairbridge, R. W. Eustatic changes in sea level, 107-110 in *Physics and Chemistry of the Earth*, volume 1, Pergamon Press (1967)

Faulkner, R. O. *The Pyramid Texts*, British Museum Press (1969)

Faulkner, R. O. *The Ancient Egyptian Book of the Dead*, British Museum Press (1972)

Foley, J. A. et al, *Feedbacks between climate and boreal forests during the Holocene epoch*, Nature, 371, 52-54 (1994)

Fong Chao, B. & Gross, R. S., *Changes in the Earth's rotation and low-degree gravitational field induced by earthquakes*, Geophys. J. R. astr. Soc. 91, 569-596 (1987)

Florensky, K. P. *Did a comet collide with the Earth in 1908?*, Sky & Telescope, 26:268-9, Nov 1963. (1963)

Frazer, Sir J. G. *Apollodorus: The Library*, William Heineman Ltd, London. (1921)

Furlong, D. *The Keys to the Temple*, Piatkus, London. (1997)

G

Gallenkamp, C. *Maya*, Third revised edition, Penguin Books, Harmondsworth. (1959)

Garwin, L. *In praise of interdisciplinarity*, Nature, 376, 547 (1995)

Gault, D. E. & Wedekind, J. A., *Experimental studies of oblique impact*, Proc. Lunar Planet Soc. Sci. Conf 9th, 3843-3875 (1978)

Gersonde, R. et al, *Geological record and reconstruction of the late Pliocene impact of the Eltanin asteroid in the Southern Ocean*, Nature, 390, 357-363 (1997)

Gill, E. D. *Age of Australite fall*, Journ. Geophys. Res. 75, pp 996-1002 (1970)

Gillespie, C. C. *Pierre-Simon Laplace (1749-1827) A life in Exact Science*, Princeton Univ. Press (1998)

Giot, P-R et al, *About the age of the oldest passage graves in western Brittany*, pp 624-6, Antiquity, 68 (1994)

Gidley, L. *Stonehenge viewed by the light of Ancient History and modern Observation*, Simpkin, Marshall & Co, London (1873)

Godwin, H. & Newton, L. *The submerged forest at Borth and Ynyslas*. New Phytologist, 37, 333-44. (1938)

Gold, T. *Instability of the Earth's Rotation*, Nature, London, 175, pp. 526-9. (1955)

Gold, T. *Irregularities in the Earth's Rotation*, Sky and Telescope, 17, 284-286 (1958)

Goslar, T. Et al, *Variations of Younger Dryas atmospheric radiocarbon explicable without ocean circulation changes*, Nature, 403, 877-879 (2000)

Greenberg, R. *The Dynamics of Uranus's Satellites*, Icarus, 24,235-322 (1975)

Greenwood, J. J. D. *Three-year cycles of lemmings and Arctic geese explained*, Nature, 328, p. 577. (1987)

Griffith, J. *The astronomical and archaeological value of the Welsh Gorsedd*, Nature, 76, 9-10. (1907)

H

Haartman, W. K., Phillips, R. T. & Taylor, G.T. *Origin of the Moon*, Lunar & Planetary Inst. Houston (1986)

Hallam, A. *Great Geological Controversies*, Oxford University Press, Oxford. (1983)

Hammer, C. U., Clausen, H. B. & Dansgard, W. *Greenland ice sheet evidence of post-glacial volcanism and its impact, Nature*, 288, 230-235 (1980)

Hammer, C.U. et al, *The Minoan eruption of Santorini in Greece dated to 1645 BC?, Nature*, 328, 517-519 (1987)

Hancock, G. *Heavens Mirror*, Michael Joseph Ltd, Penguin (1998)

Handford, S. A. *Caesar: The Conquest of Gaul*, Penguin Books, Harmondsworth. (1951)

Hapgood, C. H. *Earth's Shifting Crust*, Pantheon, New York (1958)

Hapgood, C. H. *Maps of the Ancient Sea Kings*, (revised edition) Turnstone Books, London (1979)

Hapgood, C. H. *The Path of the Pole*, Souvenir Press Ltd, London (1999)

Harris, J. R. *The Legacy of Egypt*, Clarendon Press, Oxford (1971)

Hawking, S. *The Universe in a Nutshell*, Bantam Press, London (2001)

Hawkins, G. S. *Stonehenge Decoded*, Souvenir Press Ltd, London. (1966)

Hayes, J. D., Imbrie, J 7 Shackleton, N., *Variations in the Earth's orbit: Pacemaker of the Ice Ages*, Science, 194, 4270, 1121-1132 (1976)

Heath, Sir Thomas L. *Greek Astronomy*, J. M. Dent, London (1932)

Heath, R. and Michell, J. *The Measure of Albion: The Lost Science of Prehistoric Britain,* Blustone Press, St Dogmaels, Wales (2004)

Heidel, A. *The Gilgamesh Epic & Old Testament Parallels*, Univ. Of Chicago Press, Chicago & London (1971)

Herries Davies, G. L. and Stevens, N. *The Geomorphology of the British Isles*, Ireland, Methuen & Co Ltd, London. (1978)

Heyworth, A. "Submerged forests around the British Isles: their dating & relevance as indicators of post-glacial land & sea-level changes", in Fletcher, J. M. (ed) *Dendrochronology in Europe*, pp. 279-88, Oxford: British Archaeological Reports International series 51. (1978)

Hildebrand, A. R. et al, *A possible Cretaceous-Tertiary boundary impact crater on the Yucatan peninsula, Mexico*, Geology 19, 867-871 (1991)

Hildebrand, A. R. et al, *Size and structure of the Chicxulub crater revealed by horizontal gravity gradients and cenotes*, Nature, 376, 415-417 (1995)

Hill, E., *On the possibility of changes in the Earth's axis*, Geol. Mag R. S., Vol 5, 262-6 (1878)

Hills, J. G. et al, *Tsunami generated by small asteroid impacts*, in Gehrels, T (ed), *Hazards due to Comets and Asteroids*, 779-799, Univ, Arizona Press (1994)

Hodges, M. A. *Society of Ley Hunters Members Questionnaire*, Hassocks (2001)

Hoffman, M. A. *Egypt before the Pharaohs*, Routledge and Kegan Paul PLC, London. (1980)

Hooke, R. *Lectures and discourses on earthquakes and subterraneous eruptions*, in *The Posthumous Works of Dr Robert Hooke* (R. Waller, ed), Smith & Walford, London (1705)

Hope, M, *Atlantis – Myth or Reality*, Penguin-Arkana (1991)

Hope, M, *The Ancient Wisdom of Atlantis*, Aquarian Press, London (1991)

Hopkins. A. A., *Legendary islands of the North Atlantic*, Scientific American Monthly 4, 362-363 and 4, 14-18. (1921)

Hopkins, W., *Researches in Physical Geology*, Phil. Trans. R. Soc. London, 129, 381-423 (1839)

Hough, S.S., *The oscillations of a rotating ellipsoidal shell containing fluid*, Philosophical Transactions of the Royal Society of London, 186, 469- 506 (1895)

Howard, W. R., *A warm future in the past*, Nature, 388, 418-9 (1997)

Howarth, H. H. *Recent changes in circumpolar lands*, Nature, 5: 162-163. (1871)

Howarth, H. H. *Circumpolar Land*, Nature, 5: 420-422. (1872)

Howarth, H. H. *Recent Climate Changes*, Nature, 6:24-25. (1872)

Hoyle, F. *On Stonehenge*, Heineman Educational Books, London. (1977)

Hoyle, F., *The Cosmogony of the Solar System*, University of Cardiff Press (1978)

Hsu, K. J. *When the Mediterranean dried up*, Scientific American, 227: 26-36. (1972)

Huber, P. J. *Astronomical Evidence for the Long and against the Middle and Short Chronologies*, (pp 15-17) in Åström (ed) *High, Middle or Low*, Gothenburg, (1987)

Hughes, M. K., *Ice-layer dating of eruption at Santorini*, Nature, 335, 211-2 (1988)

Hutton, J. (1788) *Theory of the Earth; or an investigation of the laws observable in the composition, dissolution, and restoration of land upon the globe*, R. Soc. Edinburgh, Trans., 1(2), 209-304 (1788)

Hutton, M. A. *The Tain*, Maunsel & Co, Dublin. (1907)

Hyde, D. *A Literary History of Ireland* (1967 edition), Ernest Benn Ltd, London. (1899)

I

Imbrie, A & Imbrie, C.P., *Ice Ages: Solving the Mystery*, Macmillan, London (1979)

J

Jacobsen, T. *The Sumerian King List - Assyriological Studies No 11*, Univ. of Chicago Press, Chicago (1939)

James, E.O. *Seasonal Feasts and Festivals*, Thames & Hudson, London. (1961)

James, P., *Centuries of Darkness, A challenge to conventional chronology of Old World archaeology*, Jonathan Cape, London (1991)

James, P., *The Sunken Kingdom*, Jonathan cape, London (1995)

Jeffreys, Sir H. *The Earth, its Origin, History & Physical Constitution*, 5th Edition, Cambridge University Press. (1970)

Jephcoat, A. & Olson, P. *Is the inner core of the Earth pure iron?*, Nature, 325, pp. 332-5. (1987)

Jewett, D., *Eyes Wide Shut*, Nature, 403, 145-8 (2000)

John, B. S. *The Ice Age: Past and Present*, Collins, London (1977)

Jones H. L. *Strabo: Geography*, Vol I, William Heineman Ltd, London. (1917)

Jones, Owen et al, *The Myvyrian Archaiology of Wales*, Second Edition, Thomas Gee, Denbigh. (1870)

Jowett, B. *The Dialogues of Plato*, Vol III, Oxford University Press, London. (1871)

K

Kapoor, R. *Chronology of Ancient India*, Genesis Publications PVT Ltd, New Delhi (1996)

Kelly-Simpson, W. (ed) *The Literature of Ancient Egypt*, with translations by R. O. Faulkner, E.F. Wente, Jr., and W.K. Simpson, Yale University Press, USA (1972)

Kidson, C. *Sea level Changes in the Holocene*, Quaternary Science Reviews, I, 121-151 (1982)

Kirby, W.F. *Kalevala; The Land of Heroes*, J.M. Dent & Co, London (1907) Republished by The Athlone Press (1985)

Klostermaier, K. K., *Hindu Writings*, Oneworld Press, Oxford (2000)

Koshar, R. *The Vedic People: Their History & Geography*, Orient Longman, New Delhi (2000)

Krupp, E. C. *In Search Of Ancient Astronomies*, Chatto & Windus, London (1979)

Krupp, E. C. *Echoes of the Ancient Skies*, Harper & Row, New York. (1983)

Kuniholm, P. I. et al, *Anatolian tree rings and the absolute chronology of the eastern Mediterranean*, 2220-718BC, Nature, 381, 780-3 (1996)

Kutzbach, J. E. & Webb, T., *Conceptual basis for understanding late-Quaternary Climates*, in Wright, H. E. (ed), Univ. Of Minnesota press, Minneapolis Press (1973)

L

Lamb, H. H. *Climate, vegetation and forest limits in early civilised times*, Phil. Trans. R. Soc. Lond. A. 276, 195-230 (1974) [195]. (1974)

Lamb, H.H. *Climate, Past, Present & Future Vol 1. Climatic History & the Future*, Methuen & Co, London. (1972)

Lamb, H. H. *Climate, Past, Present & Future Vol 2. Climatic History & the Future*, Methuen & Co, London. (1977)

Lamb, H. H., *Climate History and the Modern World*, Routledge, London & New York (1995)

Lambeck, K. *Geophysical Geodesy: The Slow Deformations of the Earth*, Clarendon Press, Oxford. (1988)

Lambeck, K. *Sea-level change and shore-line evolution in Aegean Greece since Upper Paleolithic time*, Antiquity, 70, 588-611 (1996)

Laskar, J., Joutel, F. & Roboutel, P. *Stabilisation of the Earth's Obliquity by the Moon*, Nature, 361, 615 (1993)

Laske, G. & Masters, G., *Limits on differential rotation of the inner core from an analysis of the Earth's free oscillations*, Nature, 402, 66-69 (1999)

Langdon, S, and Fotheringham, J.K. *The Venus Tablets of Ammizaduga*, Oxford Univ. Press (1928)

Lawton, I, & Ogilvie-Herald, C., *Giza: The Truth*, Virgin Publishing Ltd (2000)

Lee, D., Plato: *Timaeus and Critias*, Penguin, HarmondswOrth (1974)

Levi-Strauss, C., *The Raw and the Cooked*, Penguin Books (1969)

Levy, D. H., *The Quest for Comets* , Oxford Univ. Press (1995)

Lockyer, J. N. *Stonehenge and other British Stone Monuments Astronomically Considered*, 2nd edition, Macmillan, London. (1909)

Long, R. D., *A re-examination of the Sothic Chronology of Egypt, Orientalia*, 43:261 (1974)

Luminet, J-P. *Black Holes*, Cambridge, Univ. Press (1992)

Lunn, M., *A First Course in Mechanics*, Oxford Science Publications (1991)

Lyell, C. *Principles of Geology* (in 3 vols), John Murray, London, (1830-1833)

M

Mackie, E. W. *Astronomer priests in Iron Age Britain*, in Archaeological Texts on Supposed Astronomical Sites in Scotland, Phil. Trans. R. Soc. Lond. A. 276, 169-194. (1974)

Mackie, E. W. *Science and Society in Ancient Britain*, Elek Books Ltd, London. (1977)

MacNeill, E. *On the Notation and Calligraphy of the Calendar of Coligny*, Eriu, X, (1926-8), 1-67. (1928)

Malville, McKim, et al, *Megaliths and Neolithic astronomy in southern Egypt*, Nature, 392, 488-491 (1998)

Manabe. S & Stouffer, R.J. *Simulation of abrupt climate change induced by freshwater input to the North Atlantic Ocean*, Nature, 378, 165-7 (1995)

Mansinha, L. & Smylie, D. E., *Effect of earthquakes on the Chandler Wobble and the Secular Polar Shift*, Journal of Geophysical Research, 72, 4731-4743 (1967)

Maunder, E. W. *The Origin of the Constellations*, Observatory, 36, 329-334 (1913)

McCall, G.J. *Meteorites and their Origins*, David & Charles, Newton Abbot (1973)

McCarthy, D. *Easter principles and a fifth-century lunar cycle used in the British Isles*, JHA, xxiv (1993)

Mac Neill, E. *On the Notation and Calligraphy of the Calendar of Coligny, Eriu*, X, pp1-67 (1928)

Meeus, J., J.Brit.Astr. Assn, 92, 124-6 (1982)

Melchior, P. M. *Rotation of the Earth*, D. Reidel Publishing Co., Holland. (1972)

Melchior, P., *For a clear terminology in the polar motion investigations*, in Federov, E. P. et al (eds) Nutation and the Earth's Rotation, 17-21, IAU (1980)

Melosh, H. J. *Around and Around we Go*, Nature, 376, 386-7 (1995)

Melville, A. D. *Ovid – Metamorphoses*, Oxford Univ. Press (1986)

Michell, J., *The View over Atlantis*, (1962)

Michell, J., *A Little History of Astro-Archaeology*, Thames and Hudson, London. (1977)

Mitchell, W. A. *Ancient astronomical observations and Near Eastern Chronology*, Journal of the Ancient Chronology Forum, 3 (1990)

Mitrovica, J. X., *Going halves over Hudson Bay*, Nature, 390, 444-447 (1997)

Morgan, J. et al, *Size and Morphology of the Chicxulub Impact Crater*, Nature, 390, 472-476 (1997)

Mörner, N.-A. *The Holocene eustatic sea level problem*. Geol. en Mijnbouw, 50, pp. 699–702. (1971)

Mörner, N.-A., *Eustasy and Geoid Changes*, Journal of Geology, 64, 2, 123– 151 (1976)

Mörner, N.-A. *The Fennoscandian uplift and late Cenozoic geodynamics: geological evidence*, GeoJournal 3.3, pp. 287-318. (1979)

Mörner, N.-A. *Eustasy and geoid changes as a function of core/mantle changes*, in Morner, N.(ed) Earth Rheology, Isostasy and Eustasy; Scientific Reports of the Geodynamics Project No 49, John Wiley & Sons, Chichester. (1980)

Moore, P. D. *Blow, blow thou winter wind*, Nature, 336, 313. (1988)

Morris, J. (ed) *Nennius, Arthurian Period Sources*, Vol 8, British History and the Welsh Annals, Philimore Press, London and Chichester. (1980)

Mowat, F. *The Alban Quest*, Wiedenfeld & Nicholson, London. (1999)

Muck, O. *The Secret of Atlantis*, English edition, William Collins & Co Ltd, London. (1976)

Muir-Wood, *Shear waves show the Earth is a bit cracked*, New Scientist 21 April 1988, pp. 44-8. (1988)

Müller, M & Taub, K,. *The Gods and Symbols of Ancient Mexico and the Maya*, Thames & Hudson, London (1983)

Murnane, W. J. *The El-Amarna Boundary Stelae Project*, Univ. of Chicago, Annual Report, 13-16, (1983-4)

N

Needham, R., *Science & Civilisation in China*: Vol 2, Cambridge Univ. Press (1959)

Newell, R. S. *Stonehenge, Department of the Environment Official Handbook*, HMSO, London. (1959)

Newham, C. A. *The Astronomical Significance of Stonehenge*, Moon Publications, Shirenewton, Gwent, Wales. (1972)

Newman, W. S. et al, *Eustasy and deformation of the geoid: 1000-6000 radiocarbon years BP*, in Morner N.-A. (ed) Earth Rheology, Isostasy and Eustasy, John Wiley & Sons Ltd, Chichester. (1980)

Norbergen, R *Secrets of the Lost Races*, New English Library, London (1977)

North, F. J. *Sunken Cities*, University of Wales Press, Cardiff. (1957)

North, J. *Stonehenge - Neolithic Man and the Cosmos*, Harper Collins, London (1996)

O

O'Flaherty, W. D. *Hindu Myths: A Sourcebook*, Penguin, Harmondsworth (1975)

O'Flaherty, W. D. *The Rig Veda*, Penguin, Harmondsworth (1981)

O'Kelly, M. *Newgrange*, Thames & Hudson, London. (1983)

Oldfather, C. H. *Diodorus Siculus Library of History*, Vol I, William Heineman Ltd, London. (1933)

Oldfather, C. H. *Diodorus Siculus Library of History*, Vol II, William Heineman Ltd, London. (1935)

Oldfather, C. H. *Diodorus Siculus Library of History*, Vol III, William Heineman Ltd, London. (1939)

Olmsted, G. *The Gaulish Calendar*, Dr Rudolph Habelt GMBH, Bonn (1992)

Oppolzer, Theodor von, *Canon of Eclipses*, Dover Publications (1962) [originally published as *Kanon der Finsternissen*, Kaiserliche Akademie der Wissenschaft (1887)]

Owen, A. L. *The Famous Druids*, Oxford University Press, Oxford. (1962)

P

Paillard, D, *The timing of Pleistocene glaciations from a simple multi-state climate model*, Nature, 391, 378–381 (1998)

Pardee, D. & Swerdlow, *Not the earliest solar eclipse*, Nature, 363, 406 (1993)

Parker, R. A. *The Calendars of Ancient Egypt*, Univ. Of Chicago Press, Chicago (1950)

Patrick, J. *Midwinter Sunrise at Newgrange*, Nature, 249, pp. 517–19. (1974)

Pearson, G. W. *How to cope with calibration*, Antiquity, 61, 98–103. (1987)

Pennick, N. and Devereux, P. *Lines on the Landscape: leys and other enigmas*, Robert Hale Ltd (1989)

Poignant, R. *Oceanic and Australasian Mythology*, Newnes Books, London. (1985)

Pomeras, A. S. & Taton, R. *History of Science: Ancient & medieval Science from the beginnings to 1450,* Basic Books, New York (1963)

Pradham, S. N. *Chronology of Ancient India*, Cosmo Publications, Delhi (1996)

Press, F. & Briggs, P., *Chandler Wobble, earthquakes, rotation and geomagnetic changes*, Nature, 256, 270–273 (1975)

Prouty, W. F. *Carolina Bays & their Origin*, in Geological Society of America, Bulletin, 63:167–224. (1952)

R

Rackham, H. *Pliny: Natural History*, Vol I, William Heineman Ltd, London. (1938)

Rackham, H. *Pliny: Natural History*, Vol II, William Heineman Ltd, London. (1942)

Rackham, H, *Pliny: Natural History*, Vol VII, William Heineman Ltd, London. (1963)

Rawson, J. *Ancient China, Art and Archaeology*, British Museum Publications (1980)

Ray, T. P. *The winter solstice phenomenon at Newgrange, Ireland: accident or design?*, Nature, 337, 343–5. (1989)

Renfrew, C. *Before Civilisation*, Jonathan Cape Ltd, London (1973)

Renfrew, C. ed *British Prehistory: A New Outline*, Duckworth, London. (1974)

Renfrew, C. *Archaeology and Language – the Puzzle of Indo-European origins*, Cape/Cambridge University Press. (1987)

Renfrew, C. *Kings, Tree Rings and the Old World*, Nature, 381, 733–4 (1996)

Restelle, W. *Traditions of the Deluge*, Bibliotheca Sacra, 64, 148–167. (1907)

Richards. E. G. *Mapping Time: The Calendar and its History*, Oxford Univ. Press (1998)

Rieu, E. V. (tr), *Homer: The Odyssey*, Penguin, Harmondsworth (1946)

Ritchie, J. C., *Postglacial Vegetation of Canada*, Cambridge Univ. Press (1987)

Ritchie, J. C. & Haynes, C. V. *Holocene vegetational zonation in the eastern Sahara,* Nature, 330, 645–7. (1987)

Roberts, G. Et al, *Intertidal Holocene footprints and their archaeological significance*, Antiquity, 70, 647–51(1996)

Roberts, N. *The Holocene, An Environmental History*, Blackwell, Oxford (1989)

Robinson, T. M. *Heraclitus: Fragments, A Text and Translation with Commentary*, Univ. Of Toronto Press (????)

Rochester, M. G., Jensen, O. G. & Smylie, D. E, *A Search for the Earth's 'Nearly Diurnal Free Wobble'*, Geophys. J. R. astr. Soc 38, 349-363 (1974)

Rohl, D., *A Test of Time*, Century, London (1995)

Roland-Giot, P, et al, *About the age of the oldest passage-graves in western Brittany*, Antiquity, 68, 624-626 (1994)

Ronan, C. *The Shorter Science & Civilisation in China*, Vols 1, II & III Cambridge University Press (1984)

Roux, G. *Ancient Iraq*, George Allen & Unwin Ltd, London. (1964)

Roy, A. E. & Clarke, D. *Astronomy (second edition) Principles and Practice*, Adam Hilger Ltd, Bristol (1982)

Ruggles, C. *Stonehenge for the 1990s*, Nature, 381, 278 (1996)

Ruggles, C. & Barclay, G. *Cosmology, Calendars & Society in Neolithic Orkney*: a rejoinder to Euan MacKie, Antiquity, 74 (2000) pp62-74.

Russel, D. & Tucker, W., *Supernovae & the extinction of the Dinosaurs*, Nature, 299, 553-554 (1971)

Ryan, W. & Pitman, W. *Noah's Flood*, (1998)

S

Safronov, V. S., *Evolution of the Protoplanetary Cloud and formation of the Earth and the Planets*, NASA TTF-677 (1972)

Saggs, H. W. F. *Babylonians*, British Museum Press

Sandars, N. K. *The Epic of Gilgamesh*, Penguin Books, Harmondsworth. (1960)

Sandars, N. K. *Poems of Heaven and Hell from Ancient Mesopotamia*, Penguin Books, Harmondworth. (1971)

Sawyer, J. F. A. & Stepehson, F. R., Literary and astronomical evidence for a total eclipse of the sun observed in ancient Ugarit on 3 May 1375 BC, *Bulletin of the School of Oriental & African Studies* (1970)

Schaaf, F., *Comet of the Century*, Springer-Verlung, New York (1997)

Schulz, J. *Movement & Rhythms of the Stars*, Floris Books, Edinburgh (1986)

Schulz, P. H. & Gault, D. E., *Prolonged global catastrophes from oblique impacts*, Geol. Soc. America, Special Papers, 247, 239-261 (1990)

Schulz, P. H. & Lianza, R. E., *Recent grazing impacts on the Earth recorded in the Rio Cuarto crater field, Argentina*, Nature, 355, 234-237 (1992)

Scott-Kilvert, I. *Plutarch – The Rise and Fall of Athens: Nine Greek Lives*, Penguin Books, Harmondsworth. (1960)

Seltz, F, *Decline of the generalist: The vigour of every discipline depends on people of broad vision*, millennium essay, Nature, 403, p 483, 3 February (2000).

Severinghaus, J. P., et al, *Timing of abrupt climate change at the end of the Younger Dryas interval from thermally fractionated gases in polar ice*, Nature, 391, 141-146 (1998)

Sherley-Price, L. *Bede — A History of the English Church and People*, Penguin Books, Harmondsworth. (1955)

Skene, W. F. *Four Ancient Books of Wales*, Edmonston & Douglas, Edinburgh. (1868)

Smith, M. L., *Wobble and Nutation of the Earth*, Geophys, J. 50, 103-140 (1977)

Sollberger, E. *The Babylonian Legend of the Flood*, British Museum Publications, London. (1971)

Song, X. & Richards, P. G., *Seismological evidence for differential rotation of the Earth's inner core*, Nature, 382, 221-4 (1996)

Spence, L. *Myths of Ancient Egypt*, George Harrap & Co Ltd, London. (1915)

Spence, L. *The Problem of Atlantis*, New York. (1924)

Spence, L. *The History of Atlantis*, Citadel Press, New Jersey (1973)

Steel, D., *Rogue Asteroids and Doomsday Comets*, John Wiley, New York (1995)

Steno, N. (1669) *The Prodromus of Nicholas Steno's Dissertation concerning a Solid Body Enclosed by process of nature Within a solid* (English translation of original Latin text by J.G. Winter) Hafner, New York (1968)

Stephenson, F. R., *Historical Eclipses and Earth's Rotation*, Cambridge University Press (1997)

Stephenson, F. R. & Houlden, M. A., *Atlas of Historical Eclipse Maps: East Asia 1500 BC to AD 1900*, Cambridge Univ. Press (1986)

Svetsov, V.V. *Total ablation of the debris from the 1908 Tunguska explosion*, Nature, 383, 697-9 (1996)

T

Taylor, T. *Proclus: Commentary on the Timaeus of Plato*, London. (1820)

Thom, A. *Megalithic Sites in Britain*, Clarendon, Oxford, (1967)

Thom, A. *Prehistoric Monuments in Western Europe*, Phil. Trans. R. Soc. Lond. A. 276, 149-156. (1974)

Thom, A. & Thom, A. S. *Megalithic Remains in Britain and Brittany*, Clarendon Press, Oxford. (1978)

Thompson, J. E. S. *Maya Astronomy*, Phil. Trans. R.Soc.Lond. A, 276, 83- 98 (1974)

Thorpe, L. *Geoffrey of Monmouth - The History of the Kings of Britain*, Penguin Books, Harmondsworth. (1966)

Tooley, M. J. *Floodwaters mark sudden rise*, Nature, 342, pp. 20-1. (1989)

Toomer, G. J. (translator) *Ptolemy: The Almagest*, Duckworth (1984)

Toomre, A., *On the 'Nearly Diurnal Wobble' of the Earth*, Geophys, J. R. Astr. Soc., 38, 335-348 (1974)

Tompkins, P. *Secrets of the Great Pyramid*, Allen Lane, London, (1973)

Turville Petre, E. O. G. *Myth and Religion of the North*, Wiedenfeld & Nicholson, London. (1964)

Twisden, Rev. J.F. *On possible displacements of the Earth's axis of figure produced by elevations and depressions of the surface*, Quart. Jorn. Geol. Soc., 133, 35-48 (1878)

V

Verschuur, G.L. *Impact! : The Threat of Comets & Asteroids*, Oxford Univ. Press (1996)

Vicente, R. O., *The Earth's constitution and the Nutations*, in Federov, E. P. et al (eds) Nutation and the Earth's Rotation, 17-21, IAU (1980)

W

Waddell, W. G. *Manetho*, Loeb Classical Library, Heineman (1940)

Walker, C. B. F. *Eclipse seen at ancient Ugarit*, Nature, 338, pp. 204-5. (1989)

Wallis Budge, E. A. *Egyptian Religion*, Kegan Paul, Trench, Trubner & Co Ltd, London. (1899)

Wallis Budge, E. A. *The Book of the Dead*, Kegan Paul, Trench, Trubner & Co Ltd, London. (1899)

Waltham, Clae, *Shu Ching: Book of History, A modernised edition of the translations of James Legge*, George Allen & Unwin Ltd, London (1971)

Warlow, P. *The Reversing Earth*, J.M. Dent & Sons, London. (1982)

Watkins, A, *The Old Straight Track* (reprinted by Garnstone Press)

Watkins, A., *The Ley Hunter's Manual* (1927)

Watson, W. *China*, Thames & Hudson (1961)

Weissman, P. R., *Terrestrial impact rates for long and short period comets*, Geol. Soc. Am., Special paper 190 (1982)

Weissman, P. R., *The Big Fizzle is Coming*, Nature, 370, p 94 (1994)

Wender, D. (tr.), *Hesiod & Theognis*, Penguin, Harmondsworth (1973)

Wentz-Evans, J., *The Fairy-Faith in Celtic Countries*, H. Frowde, London, 1911 (republished by Lemma Publishing, New York, 1973)

Werner, E. T. C. *A Dictionary of Chinese Mythology*, Kelly & Walsh Ltd, Shanghai. (1932)

West, J. A.., *The Serpent in the Sky*, Quest Books, Wheaton Illinois, (1993)

West, M. L. (ed) *Hesiod's Works & Days - edited with prologomena & commentary*, Clarendon Press, Oxford. (1978)

Weyer, E. M. *Pole movement and sea levels*, Nature, 273, 18-21 (1978)

Whiston, E. M. Josephus - *The Complete Works*, Pickering & Inglis, London. (1960)

Williams Ab-Ithel, J. *Barddas Vols I & II*, Welsh Manuscript Society, Longman & Co., London (1862)

Williams, D. M. et al, *Low-latitude glaciation and rapid changes in the earth's obliquity explained by obliquity-oblateness feedback*, Nature, 396, 453-455 (1998)

Williamson, T. & Bellamy, E. *Ley Lines in Question*, Frome (1983)

Wilson, C. *From Atlantis to the Sphinx*, Virgin Books, London (1996)

Werner, A. G. (1786) *A Short Classification and Description of the various Rocks* (trans. A. Ospovat), Hafner, New York

Werner, E. T. C. *A Dictionary of Chinese Mythology*, Kelly & Walsh, Shanghai (1932)

Z

Zachner, R. C., *Hindu Scriptures*, Everyman's library (1938)

Zangger, E., *The Flood from Heaven*, Sidgewick & Jackson, London (1992)

Zatman, S & Bloxham, J., *Torsional oscillations and the magnetic field within the Earth's core*, Nature, 388, 760-3 (1997)

Internet Websites

Although these tend to be ephemeral in nature, the following sources were in existence at time of preparation of this bibliography and were influential to the research.

www.standford.edu/~meehan/donnelly/ A collection of research concerning climate and sea level changes around 3000-3200 BC

Index

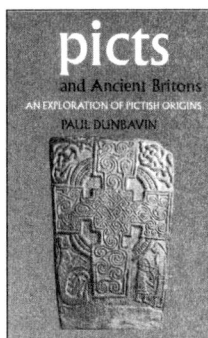

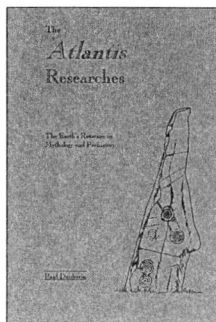